Microbial Fuel Cells

The Wiley Bicentennial–Knowledge for Generations

Each generation has its unique needs and aspirations. When Charles Wiley first opened his small printing shop in lower Manhattan in 1807, it was a generation of boundless potential searching for an identity. And we were there, helping to define a new American literary tradition. Over half a century later, in the midst of the Second Industrial Revolution, it was a generation focused on building the future. Once again, we were there, supplying the critical scientific, technical, and engineering knowledge that helped frame the world. Throughout the 20th Century, and into the new millennium, nations began to reach out beyond their own borders and a new international community was born. Wiley was there, expanding its operations around the world to enable a global exchange of ideas, opinions, and know-how.

For 200 years, Wiley has been an integral part of each generation's journey, enabling the flow of information and understanding necessary to meet their needs and fulfill their aspirations. Today, bold new technologies are changing the way we live and learn. Wiley will be there, providing you the must-have knowledge you need to imagine new worlds, new possibilities, and new opportunities.

Generations come and go, but you can always count on Wiley to provide you the knowledge you need, when and where you need it!

William J. Pesce
President and Chief Executive Officer

Peter Booth Wiley
Chairman of the Board

Microbial Fuel Cells

Bruce E. Logan
The Pennsylvania State University

WILEY-INTERSCIENCE
A John Wiley & Sons, Inc., Publication

Published by John Wiley & Sons, Inc., Hoboken, New Jersey.
Published simultaneously in Canada.

Library of Congress Cataloging-in-Publication Data:

Logan, Bruce E.
Microbial fuel cells / Bruce E. Logan.
p.; cm.
Includes index.
ISBN 978-0-470-23948-3 (cloth)
1. Biomass energy. 2. Fuel Cells. 3. Microbial biotechnology. I. Title.
TP339.L64 2007
621.31'2429—dc22

2007035520

Printed in the United States of America.

10 9 8 7 6 5 4 3 2 1

To Maggie, Alex and Angela,
for their love and support

Contents

Preface

This book is made possible by work performed in my laboratory as well as experience gained through numerous collaborations that were started only a few years ago, but the roots of my education in the area of exogenous electron transfer go back two decades. I first began to learn about iron reducing bacteria from Bob Arnold and his group in 1986, when Bob and I were both assistant professors at the University of Arizona. Bob was an early pioneer in the area of solid iron oxide reduction by bacteria, and I am grateful to Bob and his students (particularly Flynn Picardal) for sharing their work and thoughts with me over the years about these fascinating bacteria.

In the fall of 2002, I started work on microbial fuel cells (MFCs) and was fortunate to work with Hong Liu, a postdoctoral researcher with my group at that time, as she made essential creative and intellectual contributions to our laboratory's work on MFCs. Through her efforts and excitement for this topic, work in my laboratory advanced at a rapid pace and we made many interesting discoveries over a short period of time. In the years since I started working on MFCs, I have been privileged to work with a number of talented researchers, but I especially appreciate having worked with the following students and researchers in my laboratory: Shaoan Cheng, Booki Min, JungRae Kim, SangEun Oh, Jenna Heilmann Ditzig, Yi Zuo, Douglas Call, Valerie Watson, Rachel Wagner, Farzaneh Rezaei and Defeng Xing. My own research efforts have always focused on the biophysical interface, and so I have relied on collaborations with others having greater expertise in chemistry, biology, molecular biology techniques. I am grateful to Tom Mallouk at Penn State for all his patient explanations of fuel cells and electrochemistry, and the assistance of his student Ramna Ramnarayanan. I particularly appreciate collaborations with Jay Regan and his group at Penn State, as their expertise and knowledge have been absolutely essential for work that has emerged from Penn State.

Collaborations outside of Penn State have been critical for advancing the field of MFC research. In 2003 I spent a sabbatical at the University of Newcastle upon Tyne, and I benefited from conversations and work with Ian Head, Tom Curtis, Cassandro Murano, and Keith Scott. Collaborations have continued with this group through the efforts of Eileen Wu, who joined my research group for several months. I am continuing to benefit from additional collaborations with Yuri Gorby (J. Craig Venter Institute), Ken Nealson and Orianna Bretschger (University of Southern California), Tim Vogel, Jean-Michel Monier (Ecole Centrale de Lyon, France), Yujie Feng and Aijie Wang (Harbin Institute of Technology, China), Kazuya Watanabe and Shunichi Iichi (Marine Biology Institute, Japan), Kyeong-Ho Lim (Kongju National University, Korea), and many others.

This book really represents the next stage of evolution of a paper written in 2006, with co-authors Peter Aelterman, Bert Hamelers, René Rozendal, Uwe Schröder, Jurg Keller, Stefano Freguia, Willy Verstraete, and Korneel Rabaey. The chapters of this book on voltages and power are based on a section of that paper originally crafted by René and Bert, with additional contributions from Uwe and Korneel. I especially appreciate the extra effort and continued discussions with René on thermodynamics, power calculations, and MECs, and with Kornell on MFCs. Each of the chapters in the book has been improved by comments provided by a number of different colleagues, and I thank especially: René Rozendal, Ian Head, Nathan Lewis, Annemiek ter Heijne, Korneel Rabaey, Ken Nealson, Uwe Schröder, Song Jin, Jurg Keller, Lenny Tender, and Denny Parker.

I thank collectively everyone I mentioned above for their help in making a book like this possible. I look forward to continued rapid developments in the exciting areas of MFCs, MECs/BEAMRs, and bioenergy production.

BRUCE E. LOGAN

State College, Pennsylvania
September 2007

CHAPTER 1

Introduction

1.1 Energy needs

There are over six billion people on the planet with 9.4 billion projected for 2050 (*Lewis and Nocera* 2006). Fossil fuels have supported the industrialization and economic growth of countries during the past century, but it is clear that they cannot indefinitely sustain a global economy. Oil will not appreciably run out for at least 100 years or more, but demand for oil is expected to exceed production from known and anticipated oil reserves ten or twenty years from now, or within the 2015 to 2025 time frame (*Rifkin* 2002). This may seem distant to many consumers and businesses that rarely plan for more than three to five years in the future, but this is a very short time frame for society as a whole. Planning a single section of an interstate highway in a city, for example, can take ten years or more. The infrastructure changes needed to address our global energy needs will be far more extensive and will likely require changes not only to our infrastructure but also to our lifestyle. Changes will affect everything from home heating and lighting, to where we prefer to live and work and how we get there. The costs of energy and how much energy we use will come to dominate our economy and our lifestyle in the coming decades.

The total annual energy consumption in the US is ~100 quads of energy (100 quadrillion BTU = 10^{15} BTU), or 1.1×10^{15} J, which is a continuous consumption rate of 3.34 TW (1 TW = 10^{12} W). On a global scale, energy use is 13.5 TW (*Lewis and Nocera* 2006), Thus, the US uses about 25% of the world's energy despite having only 5% of the world's population. Energy in the US is derived from a number of sources, but most are fossil fuels (Fig. 1.1). Approximately 18% of this energy (600 GW) is generated as electricity at power plants that vary greatly in size, with a typical large power plant producing ~1 GW. Power plants are 33% efficient, so energy used to make this electricity is larger by a factor of three.

If we assume a base of 300 million people in the US, each person in the US consumes on average 11.1 kW, or 97 MWh per year. This is not a level of energy utilization that we see in our daily life as much of this energy is used for manufacturing and transportation or is lost as heat in various energy conversion and utilization cycles. At a more local level, an average US house uses 1.22 kW while a home in British

Columbia, Canada uses 1.5 kW (non-electric heating) to 2.5 kW (electric heating) (*Levin et al.* 2004). In comparison, 500 gallons of gas is annually used per person in the US, or an energy equivalent of 2.1 kW.

How much energy will we need in the future? One estimate of population growth, coupled economic growth at current levels, puts a global demand of 41 TW in 2050 at current energy growth rates. However, considering anticipated energy trends, a more reasonable projection is 27 TW by 2050 and 43 TW by 2100 (*Lewis and Nocera* 2006).

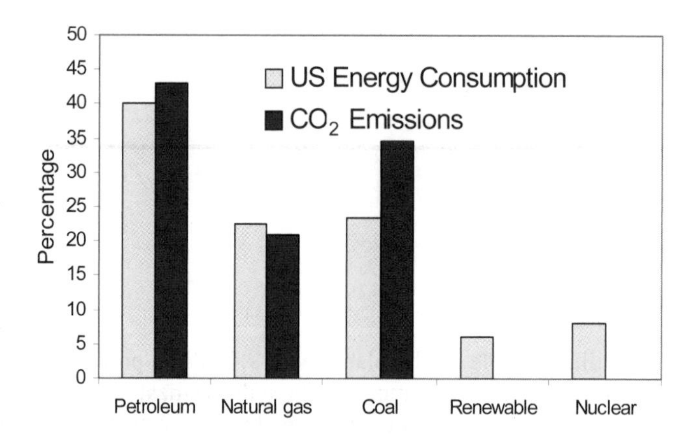

Fig. 1.1 Percent of US energy (98.5 Quad, 2003) by source and CO_2 emissions (5.772×10^9 metric tons. [From *Shinnar and Cintro* (2006). Reprinted with permission from AAAS.]

1.2 Energy and the challenge of global climate change

There is no "magic bullet" for meeting our current or future energy needs. While oil, natural gas, and coal are the main methods for energy production today (Fig. 1.1), this cannot continue into the future. When the US had its first oil crisis in the 1970s due to demand exceeding production, the solution was simply to find other sources of oil. However, finding new sources of oil, increasing the efficiency of extracting oil from existing source, or using other fossil fuels such as tar sands and shale oil will not address an equally challenging task of addressing climate change. There is no question that the release of stored carbon in fossil fuels is increasing the concentration of carbon dioxide in the atmosphere, with increases from 316 ppmv in 1959 to 377 ppmv in 2004 (Fig. 1.2A). By 2100, it is estimated that CO_2 concentrations will reach anywhere from 540 to 970 ppmv. Without substantial changes to our energy production methods, we will have greatly exceeded any historic level of CO_2 concentrations in the atmosphere. Global CO_2 accumulation over the next 40–50 yr will persist for the next 500–2,000 yr, and thus even the current levels of CO_2 are not likely to soon change. Global mean temperatures have already risen above pre-historic levels (Fig. 1.2B), resulting in melting of glaciers and rising sea levels.

The insufficient availability of oil and natural gas could be augmented by other fuels, such as coal, coal tars, oil shales, and methane hydrates. However, if we obtain energy from these sources using conventional technologies, we will release additional CO_2, exacerbate environmental damage, and accelerate global climate change. Carbon capture and sequestration could be used, but this will continue to add to releases produced by other fuels. Clearly, we would need a very efficient method of carbon sequestration. We would need to develop a method that did not leak CO_2 into the atmosphere at an average rate (globally!) of more than 1% over centuries (*Lewis and Nocera* 2006). Of course this approach requires that all countries that release CO_2 are equally committed and effective in carbon capture and sequestration.

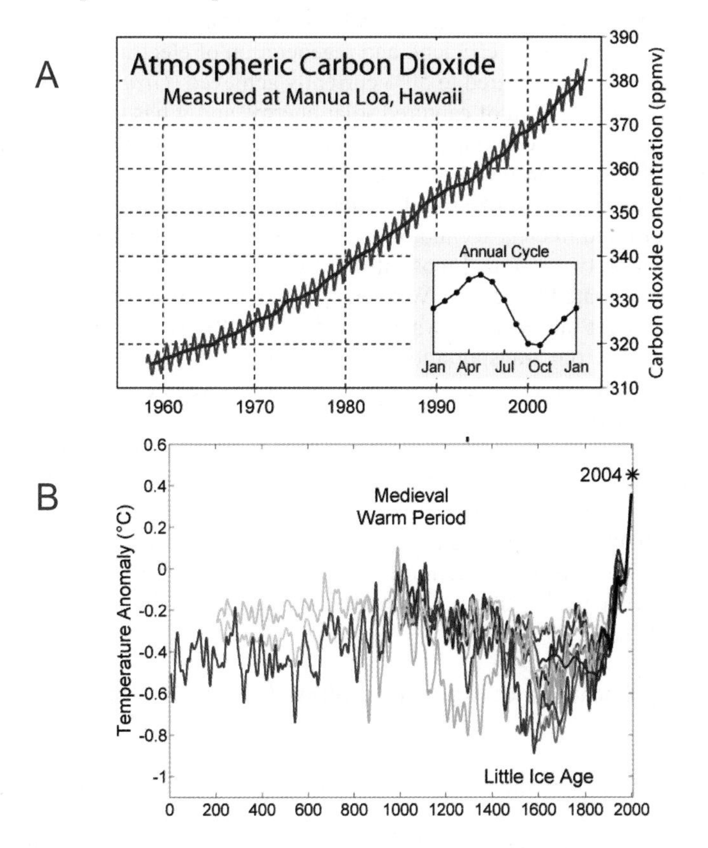

Fig. 1.2 (A) CO_2 concentrations in the atmosphere measured at Manua Loa (*Wikipedia-Contributors* 2007a); (B) temperatures relative to historic levels (*Wikipedia-Contributors* 2007b).

Our greatest environmental challenge is to simultaneously solve energy production and CO_2 releases: We must develop a whole new energy platform that produces sufficient energy while at the same time reduces CO_2 emissions. Our goal must be to meet 2050 energy needs on a carbon neutral basis.

Nuclear fission alone is not the answer. The expected availability of uranium is estimated to produce only 100 TW-h of electricity, and thus if 10 TW of power was obtained from nuclear energy the supply of uranium would be depleted in less than a decade. Moreover, we would need to build a new 1 GW plant every 1.6 days for the next 45 years (*Lewis and Nocera* 2006). This scenario doesn't even address the environmental and human damage caused by uranium mining, or the lack of a safe, long-term solution for the storage of nuclear waste.

Solar energy is ultimately the long-term solution, but it all depends on how we harvest this source of energy. We currently use in 1 hour the energy of sunlight (4.3×10^{20} J) that strikes the planet each year. The sun does not shine all day, nor does it shine equally in all regions. Thus, solar panels can help with daytime electricity needs, but it will not serve as a primary source of energy throughout the day and night without efficient methods of energy storage. Water electrolysis to produce hydrogen is a useful approach as electrolysis can be quite efficient, and regeneration of electricity in hydrogen fuel cells could approach 80% compared to 50% currently achieved (*Grant* 2003).

Biomass energy is another form of captured solar energy, and it has the advantage of solar energy being stored in the biomass for concentrated processing and delivery. Shinnar and Cintro (2006) have proposed a roadmap to switch to non-fossil fuels over the next 30 to 50 years that rely on proven and available technologies that include solar, geothermal, wind, hydro and nuclear technologies. Roughly 15% of the land area in the US is currently used for cultivation, and most of that for food (*Grant* 2003). The energy value of these crops is estimated to be 2.6 TW of energy, which clearly is not sufficient to meet existing energy needs alone (and what would we eat?). However, to produce sufficient hydrogen to meet our transportation infrastructure with crops we would need 0.4 TW to make hydrogen via water electrolysis. This is equivalent to an increase in cultivation to 18% of the land (*Grant* 2003).

Costs to replace 70% of the fossil fuels (and most coal sources) would cost $170 to $200 billion per year over the next 30 years (*Shinnar and Cintro* 2006). While such economic figures are indeed challenging, they are not insurmountable. Given this long time frame, it is also likely that new technologies could emerge that could change the economic assessment. Thus, our best solution for both energy and climate appears to be heavy investment in renewable energy resources, in terms of both research and development.

1.3 Bioelectricity generation using a microbial fuel cell—the process of electrogenesis

Microbial fuel cell (MFC) technologies represent the newest approach for generating electricity—bioelectricity generation from biomass using bacteria. While the first observation of electrical current generated by bacteria is generally credited to Potter in 1911 (*Potter* 1911), very few practical advances were achieved in this field even 55 years later (*Lewis* 1966). In the early 1990s, fuel cells became of more interest and work on MFCs began to increase (*Allen and Bennetto* 1993). However, experiments that were conducted required the use of chemical mediators, or electron shuttles, which could carry electrons from inside the cell to exogenous electrodes (see Chapter 2). The breakthrough in MFCs occurred in 1999 when it was recognized that mediators did not need to be added (*Kim et al.* 1999c; *Kim et al.* 1999d).

In an MFC, microorganisms degrade (oxidize) organic matter, producing electrons that travel through a series of respiratory enzymes in the cell and make energy for the cell in the form of ATP. The electrons are then released to a terminal electron acceptor (TEA) which accepts the electrons and becomes reduced. For example, oxygen can be reduced to water through a catalyzed reaction of the electrons with protons. Many TEAs such as oxygen, nitrate, sulfate, and others readily diffuse into the cell where they accept

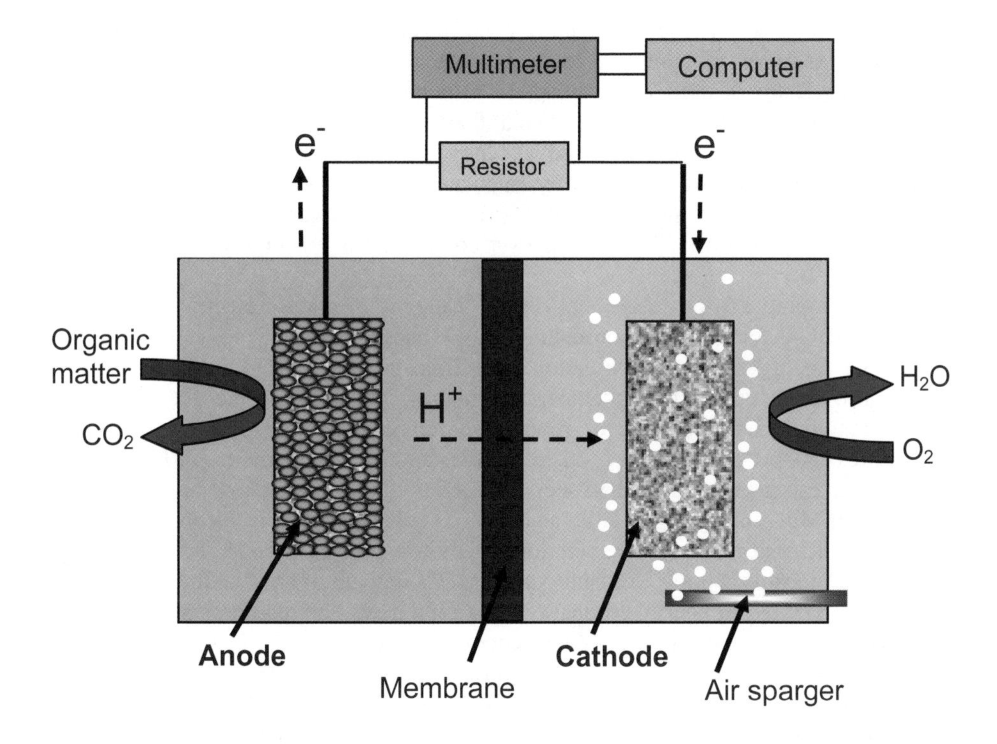

Fig. 1.3 Schematic of the basic components of a microbial fuel cell (not to scale). The anode and cathode chambers are separated by a membrane. The bacteria grow on the anode, oxidizing organic matter and releasing electrons to the anode and protons to the solution. The cathode is sparged with air to provide dissolved oxygen for the reactions of electrons, protons and oxygen at the cathode, with a wire (and load) completing the circuit and producing power. The system is shown with a resistor used as the load for the power being generated, with the current determined based on measuring the voltage drop across the resistor using a multimeter hooked up to a data acquisition system.

electrons forming products that can diffuse out of the cell. However, we now know that some bacteria can transfer electrons exogeneously (*i.e.,* outside the cell) to a TEA such as a metal oxide like iron oxide. It is these bacteria that can exogenously transfer electrons, called *exoelectrogens*, that can be used to produce power in an MFC. The nomenclature used for categorizing process, microorganisms, and reactors for methane generation is: methanogenesis, methanogens, and anaerobic digesters. Similarly, we classify this method of electron-generating process as *electrogenesis*, with the bacteria *exoelectrogens* and the reactor a *microbial fuel cell (MFC)*.

A schematic of an MFC system is shown in Fig. 1.3. Oxygen in the anode chamber will inhibit electricity generation, so the system must be designed to keep the bacteria separated from oxygen (the catholyte in this example). This separation of the bacteria from oxygen can be achieved by placing a membrane that allows charge transfer between the electrodes, forming two separate chambers: the anode chamber, where the bacteria grow; and the cathode chamber, where the electrons react with the catholyte. The cathode is sparged with air to provide dissolved oxygen for the reaction. The two electrodes are connected by a wire containing a load (*i.e.,* the device being powered), but in the laboratory a resistor is used as the load. In principle, the membrane is permeable to protons that are produced at the anode, so that they can migrate to the cathode where they can combine with electrons transferred via the wire and oxygen, forming water. The current produced by an MFC is typically calculated in the laboratory by monitoring the voltage drop across the resistor using either (a) a voltmeter (intermittent sampling) or (b) a multimeter or potentiostat hooked up to a computer for essentially continuous data acquisition.

The development of processes that can use bacteria to produce electricity represents a fantastic method for bioenergy production as the bacteria are self-replicating, and thus the catalysts for organic matter oxidation are self-sustaining. Bacterial reactions can be carried out over several different temperature ranges depending on the tolerance of the bacteria, ranging from moderate or room-level temperatures (15–35°C) to both high temperatures (50–60°C) tolerated by thermophiles and low temperatures (<15°C) where psychrophiles can grow. As we shall see, virtually any biodegradable organic matter can be used in an MFC, including volatile acids, carbohydrates, proteins, alcohols, and even relatively recalcitrant materials like cellulose.

While the idea of making electricity using MFCs may not be new in theory, certainly as a practical method of energy production it is quite new. The requirements for making MFCs economically viable as a method of energy production are demanding. The cost of oil currently remains low, and there are many different alternative methods of energy production that have reached a high level of development making them competitive for energy production. MFCs are so new that relatively little effort has been put into practical architectures using affordable materials. As highlighted in this book, however, that is already changing and many new approaches for MFC design are yielding promising results. When a new technology is developed, the fastest way to bring it to the market is to apply it in an area most likely to yield the greatest profit. As the technology further develops, it can then reach new markets. Computer hard drives needed many years of development, for example, before they could be small enough to be portable as music players. Similarly, MFCs should be developed for application in the area that will likely produce the greatest profit. For many reasons described below, it appears that the first and most useful widespread application of MFCs will be as a method of energy recovery to make the water infrastructure sustainable.

1.4 MFCs and energy sustainability of the water infrastructure

Over two billion people on the planet lack adequate sanitation, and one billion do not have sufficient access to potable water. Energy demands for conventional water and wastewater processes are a large part of the problem. In the US, we use approximately 4-5% of our electricity production for the water infrastructure, which includes water treatment and distribution, and wastewater collection and treatment. Approximately 1.5% of our electricity goes to wastewater treatment alone. The costs for maintaining the infrastructure are significant, with an annual cost for wastewater treatment of $25 billion. It is expected that over the next twenty years an additional $45 billion will need to be expended to maintain and improve this infrastructure (*WIN* 2001).

Wastewaters contain energy, in the form of biodegradable organic matter, that we expend energy to remove rather than trying to recover it. At a conventional wastewater treatment plant in Toronto, Canada, it was estimated that there was 9.3 times as much energy in the wastewater than was used to treat the wastewater (*Shizas and Bagley* 2004). Domestic, animal and food processing wastewaters are estimated to contain a total of 17 GW. This is about the same amount of energy that is currently used for the whole water infrastructure in the US (*Logan* 2004). Thus, if we could recover this energy we could make the water infrastructure self sufficient. Such an achievement would be a huge benefit to the health and well being of the US in the coming years of energy uncertainty. More importantly, such treatment processes could improve the quality of human life globally, as well as contribute to the reduction of the spread of waterborne disease through untreated sewage. Anaerobic digestion processes, based on methane generation can be an important part of energy generation from waste materials. However, they require relatively elevated temperatures (36°C) and long detention times, making them suitable only for high-strength wastewaters.

1.5 MFC technologies for wastewater treatment

Microbial fuel cell (MFC) technologies are a promising and yet completely different approach to wastewater treatment as the treatment process can become a method of capturing energy in the form of electricity or hydrogen gas, rather than a drain on electrical energy. In the late 1990s, Kim and coworkers demonstrated that bacteria could be used in a biofuel cell as a method of determining the concentration of lactate in water (*Kim et al.* 1999d), and then that electricity generation in an MFC could be sustained by starch using an industrial wastewater (*Kim et al.* 1999c). However, the power production was low and it was not clear whether the technology would have much impact on reducing wastewater strength. In 2004, this changed and the link between electricity using MFCs and wastewater treatment was clearly forged when it was demonstrated that domestic wastewater could be treated to practical levels while simultaneously generating electricity (*Liu et al.* 2004). The amount of electricity generated in this study, while low (26 mW/m^2), was considerably higher (several orders of magnitude) than had previously been obtained using wastewater. Research led by Reimers (2001) a few years earlier had demonstrated that organic and inorganic matter in marine sediments could be used in a novel type of MFC, making it apparent that a wide variety of substrates, materials, and system architectures could be used to capture electricity from organic matter with bacteria. Still, power levels in all these systems were relatively low. The final development that sparked the current interest in MFCs was provided by Rabaey *et al.* (2003) when they demonstrated power densities two orders of magnitude greater was possible in an MFC using glucose, again without the need for exogenous chemical mediators.

Following these demonstrations, the race was on to develop practical applications of MFCs, with the first goal being development of a scaleable technology for the treatment of domestic, industrial, and other types of wastewaters (*Logan et al.* 2006). While the energy that could be captured from wastewater is not enough to power a city, it is large enough to run a treatment plant. With advances, capturing this power could achieve energy sustainability of the water infrastructure. As an example of the power that can be derived from wastewater, consider the example that follows for energy recovery for a modest-sized town.

Example 1.1

What is potential energy benefit of maximum energy recovery using domestic wastewater to a town of 100,000 people? (a) Calculate the maximum energy production for assuming 500 L/d per capita, 300 mg/L of COD, and 14.7 kJ/g-COD (based on wastewater solids (*Shizas and Bagley* 2004)). (b) How much is this electricity worth at \$0.44/kW-h? (c) How many homes would this power, assuming 1.5 kW/home.

(a) We can calculate power in megawatts (MW) from a simple unit conversion calculation with the given assumptions as:

$$P = \left(300\,\frac{\text{mg COD}}{\text{L}}\right)\left(\frac{500\,\text{L}}{\text{d - cap}}\right)(10^5\,\text{cap})\frac{\text{g}}{10^3\,\text{mg}}\,\frac{14.7\,\text{kJ}}{\text{g - COD}}\,\frac{1\,\text{kWh}}{3600\,\text{kJ}}\,\frac{1\,\text{d}}{24\,\text{h}}\,\frac{\text{MW}}{10^3\,\text{kW}}$$

$$P = 2.6\,\text{MW}$$

(b) The result above is for continuous power generation, so converting it to kWh and using the given electricity cost, we calculate the value of this power as

$$\text{Value} = (2.6\,\text{MW})\left(\frac{\$0.44}{\text{kWh}}\right)\frac{10^3\,\text{kW}}{\text{MW}}\,\frac{24\times 365\,\text{h}}{\text{yr}} = \$10\times 10^6\,\text{yr}^{-1}$$

Thus, we see that the power could be worth as much as \$10 million per year, although the value of electricity varies widely across the US.

(c) The number of homes (h) served with this power is just the ratio of the power generated to electricity needed per home, or

$$h = (2.6\,\text{MW})\left(\frac{\text{homes}}{1.5\,\text{kW}}\right)\frac{10^3\,\text{kW}}{\text{MW}} = 1700\ \text{homes}$$

These calculations all assume 100% energy recovery, which as we will see later is not reasonable. As a goal, it is hoped to recover 25–50% of the energy and thus the above numbers should be reduced by one half or more to obtain more practical results.

As we can see from the above example, the amount of power that can be generated from wastewater treatment for a large number of people is substantial, although on a per-person basis the energy is not particularly impressive. For the above case, the wastes from one person could generate a maximum of only 25 W, which is the power needed for a small light bulb. For a family of four, this is 100 W or that of a large incandescent light bulb. Thus, the value of the energy that can be captured from domestic wastewater is not substantial on a single home basis, but for a large plant the energy can become significant even for relatively dilute domestic wastewater.

The most significant energy savings associated with the use of MFCs for wastewater treatment, besides electricity generation, result from savings in expenses for aeration and solids handling (see Chapter 9). The major operating costs for wastewater treatment are wastewater aeration, sludge treatment, and wastewater pumping. Aeration alone can account for half of the operation costs at a typical treatment plant. Eliminating these costs can save an appreciable amount of energy. The MFC process is inherently an anaerobic process, although, as we shall see, oxygen can diffusive into the system resulting in some aerobic organic matter removal. The sludge yields for an anaerobic process are approximately one-fifth of that for an aerobic process. Thus, using MFCs could drastically reduce solids production at a wastewater treatment plant, substantially reducing operating costs for solids handling.

The MEC process for hydrogen production. An MFC is designed to generate electricity based on electrons generated by bacteria from the oxidation of organic matter. However, instead of producing electricity, the potential generated from the electrolysis of the organic matter by the bacteria can be used to produce hydrogen. The details on how this is accomplished using a microbial electrolysis cell (MEC), also known as a bioelectrochemically assisted microbial reactor (BEAMR), are described in Chapter 8. The basic idea is that the potential generated at the anode in a typical MFC can be augmented with an additional voltage to generate hydrogen gas at the cathode. Thus, if we omit oxygen at the cathode, and add in about 0.23 V or more in practice (0.11 V in theory), we can form H_2 gas at the cathode.

The concentration of organic matter in wastewater is usually evaluated on the basis of the amount of oxygen used to oxidize organic matter, in terms of biochemical oxygen demand (BOD) in a five-day biodegradation test, or via chemical oxygen demand (COD) in a chemical test that fully oxidizes all organic matter whether it is biodegradable or not. On the basis of COD, it is easy to determine the potential for hydrogen production as one mole of COD indicates that one mole of O_2 is needed for the reaction. Thus, each mole of COD (or O_2) oxidized produces 4 electrons, or the potential for 2 moles of H_2 (1 mol-COD = 2 mol-H_2). As oxygen has a molecular weight of 32 g/mol and H_2 has a molecular weight of 2 g/mol, this means that one gram of COD produces 0.125 g-H_2.

Example 1.2

How much hydrogen could potentially be generated from the domestic wastewater for a town of 100,000 people? (a) Calculate the hydrogen in kg for one year using information in the above example. (b) What would be the worth of the hydrogen assuming \$6/kg-H_2?

(a) We can calculate mass using with the above results for one year as

$$m_{\mathrm{H2}} = \left(300 \frac{\mathrm{mg\,COD}}{\mathrm{L}}\right)\left(\frac{500\,\mathrm{L}}{\mathrm{d\text{-}cap}}\right)(10^5\,\mathrm{cap})\left(\frac{0.125\,\mathrm{g}-\mathrm{H_2}}{\mathrm{g\text{-}COD}}\right)\frac{\mathrm{kg}}{10^6\,\mathrm{mg}}\;365\,\mathrm{d}$$

$$m_{\mathrm{H2}} = 6.84 \times 10^5\ \mathrm{kg}$$

(b) The value of this hydrogen, based on the given information, is:

$$\mathrm{Value} = (6.84 \times 10^5 \frac{\mathrm{kg}}{\mathrm{yr}})(\frac{\$6}{\mathrm{kg}}) = \ \$4.10 \times 10^6\ \mathrm{yr}^{-1}$$

Thus, we see that the hydrogen could be worth \$4.10 million per year. This calculation, however, does not include the energy needed to be put into the process or costs for gas purification and compression, and it is a maximum assuming no loss of substrate to cell mass or other processes and complete recovery of the hydrogen produced.

1.6 Renewable energy generation using MFCs

In order for MFCs to compete with other technologies in renewable energy generation, the costs of making the reactors and the source of the material must in the near term be competitive with fossil fuels. If carbon taxes are included in the costs of making energy using fossil fuels, then MFCs and other renewable energy methods will become competitive. The cost to build an electric power plant is $1,000 per kW (*Grant* 2003). The costs for making an MFC reactor are unknown, as a commercial system has yet to be built. We can look to comparisons to trickling filter media, which costs ~$530 per m^3, with a total surface area of 100 m^2. Assuming we achieve 1 W/m^2, this would translate to cost for the media of $530/0.1 kW or $5300 per kW. Electrically conductive media could cost more than trickling filter media, but it may be possible to achieve much higher surface areas than with traditional TF media (see Chapter 9 on wastewater applications). This estimate also does not consider the cost of the cathode, electrical systems for DC-DC or DC-AC conversions, and other factors.

Assuming costs can be reduced for building MFCs, what are the potential applications for using biomass for energy recovery? Biomass has already surpassed hydropower as the leading source of renewable energy in the US, providing 3% (2.9 Quads per year, or 97 GW) of the energy in the US equal to 47% of the renewable energy (*Perlack et al.* 2005). Hydroelectric (45%), geothermal (5%), wind (2%), and solar (1%) form the remainder. A recent study for the US Department of Energy concluded that it is possible to obtain 1.3 billion dry tons of biomass per year, more than sufficient to replace over 30% the present petroleum consumption in the US which was estimated to require 1 billion dry tons (*Perlack et al.* 2005). The goal is to provide 5% of the nation's power, 20% of transportation fuels, and 25% of chemicals by 2030.

Example 1.3

How much energy is in the food that you eat? Determine the power generation in watts possible from a 2000-Calorie per day diet.

To answer this question, we must first realize that a "Calorie" used by nutritionists is not the same as "calorie" used for engineering calculations. One Calorie is equal to 1000 cal, or 1 kcal. Thus, a daily diet of 2000 Calories is equal to 2000 kcal. Thus, the power is

$$P = \left(\frac{2000\ \text{Cal}}{\text{d}}\right)\frac{1\ \text{kcal}}{\text{cal}}\frac{10^3\ \text{Wh}}{860\ \text{kcal}}\frac{\text{d}}{24\,\text{h}} = 97\ \text{W}$$

Thus we see that the food you eat, if converted with 100% efficiency to electricity, would only provide about ~100 W per day, or the amount of power in a large light bulb. Considering that each person uses 11.1 kW of energy as a part of their daily lifestyle, we can see that food as fuel has large challenges to sustain our society. Thus, we need to look to biomass, not just food, as energy sources.

One promising method of energy recovery from biomass is to use corn stover biomass, or the stalk, husk, and leaves remaining after the harvesting of corn. There are 250 million dry tons of corn stover produced annually (*Atchison and Hettenhaus* 2004), with 90% left unused in the fields and 150 million tons that could easily be recovered for reuse (*Glassner et al.* 1999). Energy can be biologically derived from the cellulose and hemicellulose portion (70%), although the lignin (15–20%) is thought only to be useful as energy based on chemical combustion processes. A steam explosion process can convert particulate corn stover into a liquid (hydrolysate) (*Datar et al.* 2006). Hydrogen can be generated by fermentation of the hydrolysate from the sugars, but fermentation yields can never exceed 4 mol-H_2/mol-glucose, and typically it is 2 mol/mol or lower (*Logan and Regan* 2006b). Electricity can be generated in a type of MFC based on fermentation of cellulose to produce hydrogen, with the hydrogen reacting on a Pt catalyst (*Niessen et al.* 2005). However, electricity can be directly generated in an MFC using corn stover hydrolysates (*Zuo et al.* 2005). Ongoing work at Penn State has further demonstrated direct electricity generation from cellulose in an MFC. By analogy, any process that produces electricity in an MFC should work for hydrogen production in a BEAMR process. Assuming 70% of the corn stover is recovered as sugar, this represents at potential energy source of 52 GW, although current energy recovery is only 10% to date (*Zuo et al.* 2005).

1.7 Other applications of MFC technologies

MFCs may have other applications in the future besides wastewater treatment and renewable energy. By emplacing the anode electrode in marine sediments and emplacing the cathode in the overlying water, it is possible to generate electricity from the bacterial decomposition of the organic matter in the sediment. There is not sufficient electricity generated to make it economically feasible as a source of renewable energy, but it could be sufficient for powering devices in remote marine and estuarine locations. These devices, known as sediment fuel cells, are discussed in greater detail in Chapter 10. It is also possible that MFCs can be modified and used as a method of bioremediation. Although this application is far less developed than other applications, so far it has been shown that MFC-based technologies could be used to remove nitrate (conversion to nitrite) and U [conversion from soluble U(VI) to insoluble U(IV)] from water. This application is also further addressed in Chapter 10.

CHAPTER 2

Exoelectrogens

2.1 Introduction

The first billion years of life on the planet evolved without the benefit of an oxic atmosphere. Anaerobes evolved over millions of years during this time using various methods to reduce compounds to support metabolism, all without gaseous oxygen to drive respiration. Bacteria that evolved were likely able to use a number of different types of electron acceptors, but the most fascinating for those of us interested in MFCs were bacteria able to transfer electrons outside the cell. We call such bacteria *exoelectrogens*, "exo-" for exocellular and "electrogens" based on the ability to directly transfer electrons to a chemical or material that is not the immediate electron acceptor. Many anaerobes can only transfer electrons to soluble compounds such as nitrate or sulfate (not cell synthesized) that can diffuse across the cell membrane and into the cell. Exoelectrogenic bacteria are distinguished from these anaerobes by their ability to directly transport electrons outside of the cell which permits them to function in an MFC.

Electrochemically active biofilms have great importance in the natural environment, principally in metal oxidation and reduction and the associated effects on mineral dissolution, the carbon cycle, and the sorption and complexation of phosphorus and heavy metals. We are also now seeing that they may have a larger role in fulfilling a need for bioenergy production through direct electricity generation.

The diversity of bacteria capable of exoelectrogenic activity is just beginning to be discovered. A tremendous amount of information has recently been obtained studying exoelectrogens from two dissimilatory metal reducing genera (*Shewanella* and *Geobacter*). The availability of genome sequences for several isolates provides and excellent opportunity to learn more about the fundamental nature of electrogenesis (*Heidelberg et al.* 2002; *Methe et al.* 2003). However, community analysis of electrochemically active biofilms in MFCs suggests a far greater diversity of exoelectrogens in these biofilms than was previously suspected. Likewise, the mechanisms of electron transfer to extracellular electron acceptors are poorly understood. Despite genetic studies with these two genera aimed at the identification of electron carriers for iron respiration, the path of electron flow from membrane to the mineral

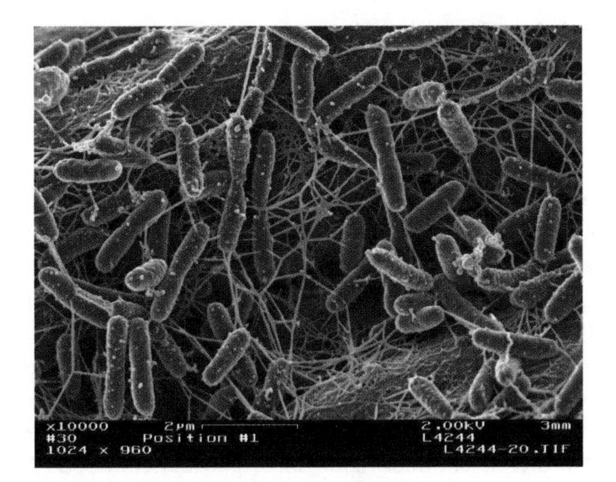

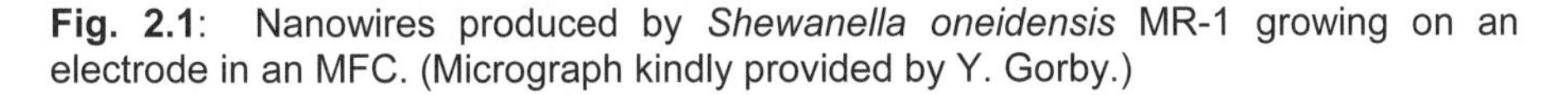

Fig. 2.1: Nanowires produced by *Shewanella oneidensis* MR-1 growing on an electrode in an MFC. (Micrograph kindly provided by Y. Gorby.)

surface continues to be debated (*Beliaev et al.* 2001; *Myers and Myers* 1998; *Myers and Myers* 2002)

Exogenous mediators have always been needed in biofuel cell experiments with *Escherichia coli* [*e.g., Sell et al.* (1989); *Park and Zeikus* (2000); *Park et al.* (2000)]. Indeed, in early experiments demonstrating the exoelectrogenic activity of *Shewanella putrefaciens*, Kim *et al.* (2002) used *E. coli* NCIB 10772 as a "control" to show the lack of electrogenic activity by this microorganism in the absence of added mediators. However, there was recently a report that power could be produced with *E. coli* K12 HB101 in an air-cathode MFC when this bacterium was "electrochemically-evolved in fuel cell environments through natural selection" (*Zhang et al.* 2006b). The researchers pumped a suspension of this bacterium into an MFC, and then obtained samples from this reactor and re-cultivated these cells. After several successive iterations, it was found that that there was increased power output in this reactor without the need for addition of mediators. This result suggests that *E. coli* can be evolved to exhibit electrogenic activity. However, this result has not been independently verified and no analysis was conducted to ensure that the reactor remained a pure culture of *E. coli.*

2.2 Mechanisms of electron transfer

Bacteria are so far *known* to transfer electrons to a surface via two mechanisms: electron shuttling via self-produced mediators (such as pycocyanin and related compounds produced by *Pseudomonas aeruginosa (Rabaey et al. 2005a; Rabaey et al. 2004)* and nanowires produced by both *Geobacter* and *Shewanella* species (*Gorby and Beveridge* 2005; *Reguera et al.* 2005). In addition, research shows that ferric iron reduction by *Shewanella* involves membrane-bound electron carriers. A number of proteins in the cytoplasmic membrane, periplasm, and outer membrane that are involved in dissimilatory mineral reduction have been identified by mutagenesis (*Beliaev and Saffarini* 1998; *Myers and Myers* 2000; *Myers and Myers* 2001) and biochemical studies (*Dobbin et al.* 1995; *Lower et al.* 2001). However, such information on electron transfer mechanisms is insufficient to describe how bacteria colonize and maintain viable cells at a surface, and competition among bacteria for the surface has not yet been examined.

2.2.1 Nanowires

Gorby and co-workers reported the occurrence of conductive appendages for both *Geobacter* and *Shewanella* species (*Gorby and Beveridge* 2005) which were termed bacterial "nanowires" (Fig. 2.1). The conductivity of the appendages was examined using conductive scanning tunneling microscopy (STM) where a sample is loaded onto a highly ordered pyrolytic graphite surface (a very flat and conductive surface), and a conductive (Pt-Ir) tip is rastered across the sample under constant-current imaging conditions. The resulting current–voltage curves demonstrate the portion of the scan that is conductive relative to the surface. As shown in Fig. 2.2, when the tip moved across the surface of the appendage, the current increased, demonstrating electrical conductivity in the *z*-plane (*i.e.*, from the tip to the surface). Based on this observation, it was concluded that these appendages were also conductive in the *x-y* plane (*i.e.*, between the cell and a surface), and thus could function as nanowires carrying electrons from the cell to a surface. Gorby *et al.* (2006) showed that mutants lacking key respiratory cytochromes assumed to be critical for electron transfer outside the cell (mtrC and omcA) produced appendages, but

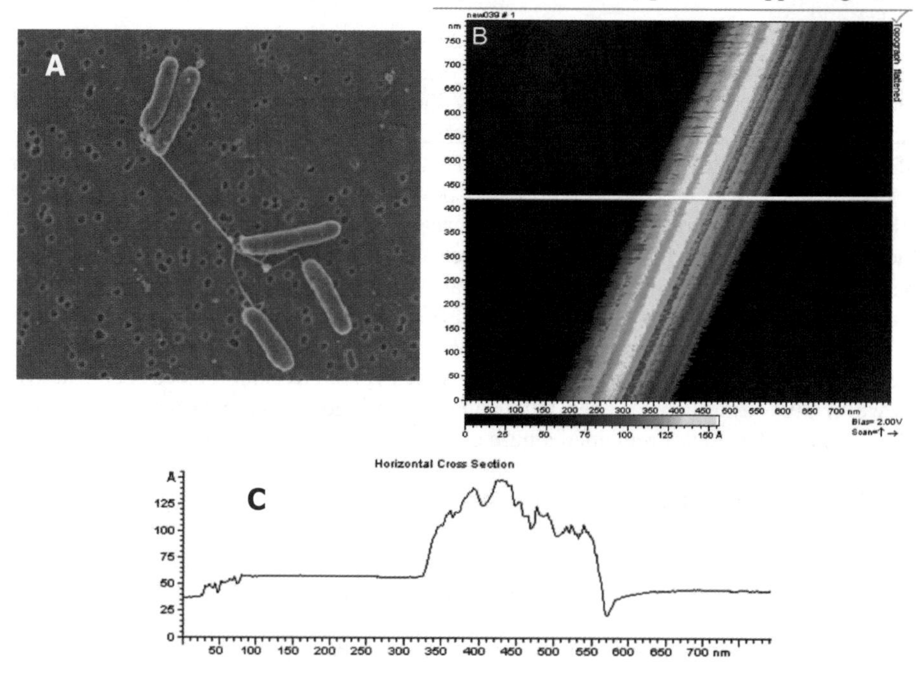

Fig. 2.2 (A) *Shewanella oneidensis* MR-1 on a filter imaged using scanning electron microscopy (SEM) showing pilus-type nanowires connecting cells. (B) Single conductive appendage from MR-1 where the white line is the electrical conductivity based on an STM scan shown in (C). Note that the multiple ridges and troughs (lateral diameter of 100 nm, topographic height of 5 to 10 nm) along the structure suggested bundled nanowires. [From *Gorby et al.* (2006), reprinted with permission of the National Academy of Sciences, U.S.A.]

that these appendages were not conductive in STM scans. Moreover, these mutants were also impaired in their ability to reduce iron or produce electricity in an MFC.

This observation of the production of conductive appendages by *G. sulfurreducens* was similarly observed by Reguera *et al.* (2005), but the structure of the nanowires produced by *G. sulfurreducens* appears to be quite different than those made by *S. oneidensis*. The appendages produced by *G. sulfurreducens* look to be relatively thin single strands (Fig. 2.3), while those produced by *S. oneidensis* have the appearance of thick "cables" that might possibly consist of several conductive wires bundled together. Note that Reguera *et al.* used conductive AFM to obtain their data, while Gorby *et al.* (2006) used STM. Reguera *et al.* concluded that appendages produced by *S. oneidensis* MR-1 were not conductive. However, examination of their supplemental material shows that they applied a voltage of only 100 mV when scanning appendages of *S. oneidensis*, while they applied ±600 mV when analyzing *G. sulfurreducens* (Fig. 2.3D). The low voltage may have been insufficient to distinguish the conductivity from the inherent "noise" based on data in Fig. 2.3D.

Evidence of nanowires produced by photosynthetic microorganisms. The production of conductive nanowires has been shown by microorganisms other than iron–reducing bacteria. The capability to produce conductive appendages has also been demonstrated in

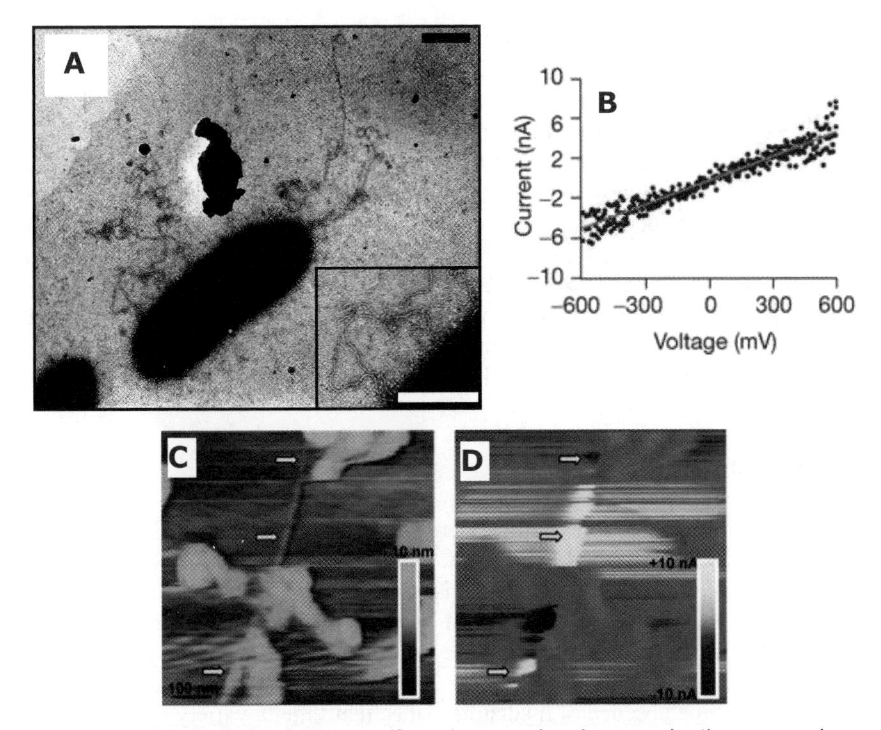

Fig. 2.3 (A) SEM of *Geobacter sulfurreducens* showing conductive appendages when grown on insoluble iron. (B) Correlation of applied voltage and current using conductive AFM. (C) Topography and (D) conductivity of the pilus (shown by arrows) in conductive scans. Note that there appears to be only one single wire in a filament here, versus the many ridges and troughs observed with S. oneidensis MR-1. [Reprinted from *Reguera et al.* (2005), with permission, Nature Publishing Group.]

phototrophic, oxygenic cyanobacteria (*Synechocystis*) (*Gorby and Beveridge* 2005). Subsequent examination of these cultures in MFCs demonstrates that when these cells are grown under CO_2 limited conditions, that they can produce electricity in an MFC in the light, but not in the dark.

The potential for interspecies electron transfer. There is also emerging evidence for interspecies electron transfer. It has been observed that *Pelotomaculum thermopropionicum* produces thick, conductive appendages (using STM) that resemble pilli (*Gorby et al.* 2006). In co-cultures, these appendages have been observed to connect these fermentative bacteria to a methanogen (*M. thermoautotrophicus*). Such a connection could facilitate electron transfer between the fermentative bacterium, which needs to release electrons in order to regenerate intracellular NADH (see Chapter 3). It is usually found that fermentative bacteria release electrons as hydrogen gas, or through the production of more reduced species, but this finding of conductive appendages raises the possibility of direct electron transfer from one microbe to another. Additional preliminary evidence collected in Gorby's lab suggests that there may also be syntrophic methanogenic/sulfate–reducing co-cultures that produce nanowires and participate in interspecies electron transfer. Clearly, more information is needed on this very interesting subject of interspecies electron transfer.

2.2.2 Cell-surface electron transfer

The presence of conductive nanowires does not exclude the possibility that these same bacteria can accomplish electron transfer from the surface of the cell to iron or an anode without long nanowires. Close examination of the micrograph in Fig. 2.1 shows that there are surface blebs, *i.e.,* protrusions on the surface that do not exist as nanowires but certainly could be conductive points of contact. Still, the small proteins responsible for electron transfer from the cell surface might not even be visible in such micrographs. Anaerobically grown *Shewanella oneidensis* adhered to an iron (goethite) surface with two to five times greater force than aerobically grown cells, and so the observation that this strain was more adhesive under anaerobic conditions might allow closer contact required for electron transfer from cell bound cytochromes even in the absence of nanowires (*Lower et al.* 2001).

2.2.3 Mediators

Chemical mediators or electron shuttles were routinely added to MFCs that resulted in electron transfer by bacteria and even yeast. In the earliest studies by Potter (1911) the yeast *Saccharomyces cerevisae* and bacteria such as *Bacillus coli* (later classified as *Escherichia coli)* were shown to produce a voltage, resulting in electricity generation. How that worked is not well known as there were no known mediators added to the cell suspensions, and *E. coli* (*Bond and Lovley* 2003) and yeast are not known to produce electricity today in the absence of mediators. Since that time, a variety of chemicals have been used to facilitate the shuttling of electrons from inside the cell to electrodes outside the cell. These exogenous mediators include, for example, neutral red (*Park et al.* 1999), anthraquinone-2-6,disulfonate (AQDS), thionin, potassium ferricyanide (*Bond et al.* 2002), methyl viologen, and others (*Logan* 2004; *Rabaey and Verstraete* 2005).

Rabaey and coworkers demonstrated that exogenous mediators did not have to be added to a culture (*Rabaey et al.* 2005a; *Rabaey et al.* 2004). These self-produced or *endogenous* chemical mediators, for example pyocyanin (Fig. 2.4) and related compounds produced by *Pseudomonas aeruginosa*, can shuttle electrons to an electrode and produce electricity in an MFC (*Rabaey et al.* 2004). The production of high concentrations of mediators by mixed cultures primarily containing *P. aeruginosa*, coupled with a very low internal resistance MFC achieved by using ferricyanide as a catholyte (instead of oxygen), produced 3.1 to 4.2 W/m^2 in MFCs (*Rabaey et al.* 2004; *Rabaey et al.* 2003). In a continuous-flow system, or a fed-batch reactor where the solution is replaced after each cycle of power generation when the substrate is exhausted, soluble mediators would be removed making it difficult for these compounds to accumulate to high concentrations. However, Rabaey *et al.* (2003) added substrate (glucose) to the reactor but did not replace the reactor solution. This allowed the mediators produced by the community to accumulate to high concentrations, and it resulted in reactor solutions that had a characteristic blue or blue/green color. They were able to show that the chemical mediators in solutions from the reactor had characteristics like those of pyocyanin produced by *Pseudomonas aeruginosa*. Community analysis of the reactor showed the presence of *P. aeruginosa* as well as several strains noted for hydrogen production.

It is not completely clear if pyocyanin compounds are produced primarily as a mechanism for exogenous electron transfer or for other reasons. These compounds are also known to have characteristics of antibiotics, and thus a main reason for excretion of these compounds may be as respiratory inhibitors or toxins to inactivate competitors (*Hernandez et al.* 2004; *Voggu et al.* 2006).

Mediator production was long suspected to be the main route of electron transfer for *S. oneidensis*, but there has been substantial controversy on this issue (*Myers and Myers* 2004; *Newman and Kolter* 2000). One of the main studies that supported mediator production by *Shewanella* was based on an observation that cells could reduce iron encapsulated within a porous silica bead (*Lies et al.* 2005). However, electron mediators that could accomplish this have never been identified. The subsequent finding that *S. oneidensis* produces nanowires raises doubts that mediators are needed to reach the iron. When this bacterium is grown in electron-acceptor limited conditions in chemostats, it is observed that cells rapidly extrude nanowires (*Gorby et al.* 2006). Furthermore, it has been shown that nanowires can penetrate into the pores of porous silica beads, thus making it possible that iron could be reduced at a distance even in the absence of mediators (Y. Gorby, *pers. comm.*). It is possible that cell damage could release chemicals that act as mediators in cell-free samples, but it now appears likely that such mediator production by *S. oneidensis* does not play a major role of particulate iron reduction or electricity generation in MFCs by this bacterial strain.

Other self-produced compounds for exocellular electron transfer. In addition to electron transfer by nanowires and endogenous mediators like pyocyanin, electron transfer can also occur through interspecies hydrogen transfer, the production of intermediate metabolites such as formate and acetate and other chemicals. An overall review of this subject is provided by Stams *et al.* (2006). However, here the focus is on direct electron transfer and thus these other routes of interspecies chemical transfer will not be addressed.

2.3 MFC studies using known exoelectrogenic strains

Few exoelectrogens have been isolated from MFCs, and therefore much of our knowledge about the properties of exoelectrogens is based on our study of existing bacterial strains that have been shown to produce electricity in MFCs. There are few isolates that have been obtained from an MFC, with most of these obtained by plating using iron as an electron acceptor. So far, there does not appear to be an isolate capable of producing power densities that are as large as those produced by mixed cultures in the same MFC. However, there are few direct comparisons of pure and mixed cultures in the same MFC, under the exact same conditions. In addition, high internal resistances of the MFCs used by many researchers examining power generation by pure cultures may not allow for very high power densities (*i.e.,* the maximum power is limited by the system, not the bacterial strain). It is therefore not clear if we have yet to isolate strains capable of high levels of power generation or if the cells must be grown in the presence of other bacteria to attain high power densities.

2.3.1 Exoelectrogens that produce electricity in the absence of exogenous mediators

Gammaroteobacteria and Shewanella spp. Kim *et al.* (1999d) first demonstrated electricity production by a bacterium in the absence of an exogenous mediator. They observed current generation by *Shewanella putrefacians* IR-1 (*Gammaproteobacteria*) in a type of reactor designed to be a lactate biosensor. The power output was quite low (0.01 mW/m^2). Although no exogenous mediators were added, cyclic voltammetry indicated of electrodes in a poised electrode MFC indicated oxidation and reduction peaks suggesting the presence of mediators (*Kim et al.* 1999b). MFC tests with this strain showed that current generation was inactivated by oxygen, but not by nitrate (*Kim et al.* 1999a).

Park and Zeikus (2002) measured a maximum power density of 10.2 mW/m^2 (Coulombic efficiency of 4%) using *Shewanella putrefacians* and a Mn^{4+}-graphite electrode and air-cathode using lactate as a substrate, and 9.4 mW/m^2 with pyruvate. Little power was generated from acetate or glucose (1.6 and 1.9 mW/m^2), consistent with the observation of a lack of substrate utilization by this microorganism. Incorporation of mediators into the graphite electrode increased power output by 10-fold. The maximum power density with *S. putrefacians* was six-fold lower than with a sewage sludge inoculum in the same device (*Park and Zeikus* 2003).

A culture of *Shewanella oneidensis* DSP10 was grown in a culture bottle on lactate, and then the cell suspension was fed into a small MFC (1.2 cm^3) using ferricyanide catholytes (*Ringeisen et al.* 2006). Power production was quite large (3 W/m^2, 500 W/m^3) when expressed solely on the basis of the MFC, *i.e.,* neglecting the volume of the culture bottle. The researchers did not compare these power densities using other strains or mixed cultures, nor could they account for the mechanism of high power generation. The addition of mediators increased power output by 30–100%.

The Gram-negative, iron-reducing bacterium *Aeromonas hydrophila* was isolated using conventional plating techniques (with ferric citrate or ferric pyrophosphate) from an MFC inoculated with an unspecified source and fed acetate (*Pham et al.* 2003). This facultative anaerobic bacterium produced power with yeast extract, but not with acetate. Electrochemical activity was observed with a medium containing Fe(III), but it was not observed when iron was omitted, suggesting that regulation of electrochemical activity

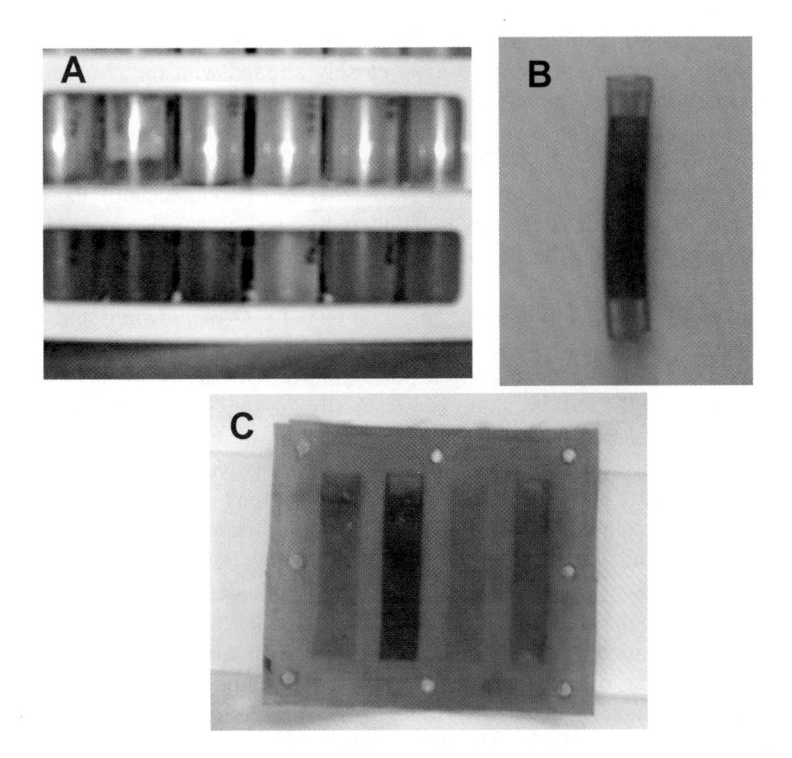

Fig. 2.4 Mediators can achieve high power densities in MFCs. (A) Solvent extracts from an MFC showing characteristic blue-green colors, typical of phenazines such as pyocyanin and phenazine-1-carboxamide produced by bacteria such as *Pseudomonas aeruginosa*; (B) a recirculation tube stained blue, likely by adsorbed mediators; (C) cation exchange membrane of 4 anodic chambers containing *P. aeruginosa* that produced different phenazines; the mediators partially sorbed onto the membrane. [Photographs by K. Rabaey.]

was linked to the presence of iron. *A. hydrophilia* KCTC 2358 also showed electrochemical activity based on cyclic voltammetry scans, suggesting that electrochemical activity might be a general property of *A. hydrophilia*.

Deltaproteobacteria and members of the Geobacteracae family. The first investigations of power generation by members of the Geobacteraceae family (*Deltaproteobacteria*) were stimulated by the observation that the majority (70%) of sequences obtained from a marine sediment MFC were most closely related to the genus *Desulfuromonas* (*Bond et al.* 2002; *Reimers et al.* 2001). The examination of power generation by a pure culture of *D. acetoxidans* in a two-chamber air-cathode MFC produced 14 mW/m^2, which was noted to be similar to that observed in the sediment MFC (16 mW/m^2) (*Bond et al.* 2002). However, direct comparison of the power densities cannot be made as it was not shown whether these devices had similar internal resistances (see Chapter 4). Current recovery with the air-cathode was high, with an 84% Coulombic efficiency. The addition of an electron shuttle (AQDS) increased power by only 24%, suggesting that power generation was near the limit for this device.

Deltaproteobacterial sulfate reducers have been shown to be abundant in sediment fuel cells, with the sequences obtained most closely aligned with the *Desulfobulbaceae* family. *Desulfobulbus propionicus* can respire using Fe(III) or AQDS and can make electricity in an MFC using pyruvate, lactate, propionate, hydrogen, or S^0, but not acetate, with the anode poised at 0.52 V (vs. NHE) (*Holmes et al.* 2004a).

Power generation was also shown for *D. acetoxidans* (using acetate) and *Geobacter metallireducens* (using benzoate) in two-chambered poised-potential system (*Bond et al.* 2002). In this case oxygen at the cathode chamber can be omitted as a potentiostat is used to poise the voltage, generating hydrogen in the cathode chamber. The recovery of electrons as current was high in both studies, with 82% for *D. acetoxidans* and 84% for *G. metallireducens*.

The maximum power generated by a pure culture of *G. metallireducens* in a two-chamber air-cathode MFC was found to be essentially the same as that of a mixed culture (38 ± 1 mW/m^2) developed from a wastewater inoculum in the same device (*Min et al.* 2005a). Maximum power densities with *G. metallireducens* were 40 ± 1 mW/m^2 in tests with ferric citrate and L-cysteine (used to scavenge oxygen) in the medium, 37.2 ± 0.2 mW/m^2 with no ferric citrate, and 36 ± 1 mW/m^2 without ferric citrate or cysteine. These small differences in maximum power could be due to slight differences in internal resistance between tests produced by using different cathodes. While cysteine was not needed for continuous power generation, its use reduced lag times for power generation likely as a result of oxygen scavenging which helped to maintain anoxic conditions. There was no evidence of cell growth supported by cysteine oxidation in pure culture tests, although cysteine has been shown to serve as sole electron donor in mixed culture MFC tests (*Logan et al.* 2005). These tests were not sufficient to determine if *G. metallireducens* could produce as much power as mixed cultures because the internal resistance of the system was very large (R_{int} = 1286 Ω). Power densities are also not a function of substrates used in this two-chamber system, as shown by similar power densities produced with ethanol or cysteine in this system (*Kim et al.* 2007c; *Logan et al.* 2005). Thus the internal resistance limited power production, not the bacteria. In tests with air-cathode MFCs having lower ohmic resistances, we can observe differences in power generation due to different substrates (*Liu et al.* 2005b) or bacteria (*Logan et al.* 2007). No effort was made to confirm the purity of the cultures following tests with *G. metallireducens*, but this should be done in future studies to ensure that the system does not become contaminated over time.

G. sulfurreducens was demonstrated to produce power (49 mW/m^2) from acetate degradation in a two-chamber MFC with an air cathode (*Bond and Lovley* 2003). Both experiments with the air cathode, and with a poised working electrode (anode) at 200 mV (vs. Ag/AgCl) showed an average electron recovery of 95%. In one case, a value of 96.8% was obtained for current recovery from oxidation of the acetate, indicating the microorganisms produced very little biomass (see Chapter 7 for a detailed discussion of cell yield). The role of hydrogen in these MFC studies is not entirely clear. Hydrogen was evolved at the cathode in poised electrode studies, and while hydrogen did not react with a sterile anode, cells metabolized hydrogen and produced current densities similar to those with acetate. Hydrogen gas crossover from the cathode to the anode chamber could have also complicated interpretation of these data. No data were provided using mixed cultures with this apparatus.

Two psychrotolerant strains isolated from the anode of a sediment MFC were shown to produce electricity and could grow at temperatures as low as 4°C to 30°C with an optimum of 22°C (*Holmes et al.* 2004c). Cells were tested in two-chamber MFCs with a poised working electrode. These isolates were classified as a new genus and species, *Geopsychrobacter electrodiphilus*, as strains A1^T and A2. They used acetate, other organic acids, amino acids, long-chain fatty acids, and aromatic compounds with growth on Fe(III) nitrilotriacetic acid as an electron acceptor.

Betaproteobacteria. Rhodoferax ferrireducens is an iron-reducing bacterium that was isolated from subsurface sediments, and shown to produce power in MFCs (*Chaudhuri and Lovley* 2003). Using glucose as an electron donor, 6.8 to 33 mW/m^2 (graphite rods, carbon felt or foam electrodes) was produced with a two-chamber system. Coulombic efficiency was 83% with a poised potential working electrode and was 81% using ferricyanide at the cathode. Fructose, sucrose, and xylose also produced power. No mixed cultures were examined in the apparatus for comparison of power densities.

Firmicutes and Clostridia. A novel Fe(III)-reducing, Gram positive bacterium was isolated from an MFC using conventional plating with ferric pyrophosphate (*Park et al.* 2001). The isolate EG3 is an obligate anaerobe most closely related to *Clostridium butyricum*, with growth at pH from 5.5 to 7.4 and at temperatures of 15–42°C. Electricity generation by EG3 (19 mW/m^2) was verified with glucose as an electron donor. The cells were shown to grow in a ferric citrate medium with cellobiose, fructose, glycerol, starch, and sucrose, among others.

2.3.2 Mediator-producing exoelectrogens

There have only been a few studies on electricity generation using mediator-producing exoelectrogens. Several mediator-producing isolates were obtained by plating from an MFC producing a high power density of 4310 mW/m^2 (*Rabaey et al.* 2004). Electrochemical activity of several isolates was shown to be due to excreted redox mediators, primarily pyocyanin produced by *Pseudomonas aeruginosa*. Power production by the isolates was much less than that produced by the mixed culture, with 28 mW/m^2 for isolate KRA3, 23 mW/m^2 for *P. aeruginosa*, and 4.9 mW/m^2 for isolate KRA1.

Additional experiments were conducted with strain KRP1 isolated from the same MFC (*Rabaey et al.* 2005a). When pyocyanin was added to cells producing electricity in a different MFC (with rod-shaped anodes), the isolate increased power generation from 0.12–0.67 mW/m^2 to 1.2–2.7 mW/m^2. Only 50% of the added pyocyanin could be recovered, suggesting that it strongly sorbed to cells and associated biopolymers. A mutant derivative of KRP1 (KRP1-phzM) that was deficient in production of pyocyanin produced much less power (0.077 mW/m^2) than the wild type, but peak power was increased with the addition of pyocyanin to 0.41 mW/m^2, with an average increase of 0.040–0.095 mW/m^2. It was also shown that addition of pyocyanin increased power by *L. amylovorus* LM1 and *Enterococcus faecium* KRA3, but decreased power output by *E. coli* ATCC4187. This demonstrated that pyocyanin can be used by other bacteria as a mediator and also that it can be inhibitory to other bacteria.

While most *Geobacter* spp. are not known to produce mediators, there is evidence for mediator production by *Geothrix fermentans* based on MFC tests using poised potential anodes (0.2 V vs. Ag/AgCl). When the medium was replaced in an MFC that had stable power generation with this isolate, power dropped by 50% and required 10 days to resume the original level. This observation was in contrast to experiments with *Geobacter sulfurreducens* which immediately produced comparable power levels with fresh medium. Filtrate from the anode chamber suspension with *G. fermentans* (obtained from a 35-day-old suspension) stimulated Fe(III) reduction by *G. sulfurreducens* better than the addition of 25 μM AQDS. Overall power production by *G. fermentans* was also indicated to be lower than that of *G. sulfurreducens*, although polarization and power density curves were not provided. The mediators thought to be responsible for these results were also not isolated or characterized by cyclic voltammetry.

2.4 Community analysis

The composition of bacterial communities that can be sustained in an MFC, either through being electrochemically active or through interactions with other bacteria, are just beginning to be studied. Molecular characterization of MFC biofilm communities to date shows that our knowledge of electrochemically active bacteria, and their interactions in biofilms, remains inadequate. While there are systems in which *Geobacter* (*Holmes et al.* 2004b) or *Shewanella* (*Logan et al.* 2005) are a predominant community member, several reports document a much broader and phenotypically uncharacterized community diversity. It is therefore clear that there is a great diversity of bacteria able to be sustained in MFCs even in cases where the biofilm is enriched through successive transfer and enrichment (*Logan and Regan* 2006a).

Even if iron-reducing bacteria are major contributors to power generation in MFCs, there still remains much to be discovered about these bacteria. Many MFCs are inoculated using untreated wastewater, solutions from wastewater treatment reactors, or wastewater treatment system sludge. In one study, microautoradiography (MAR) using radiolabeled acetate was used with ferric iron as the sole electron acceptor to examine the abundance of iron-reducing bacteria in an activated sludge reactor used for wastewater treatment. Sulfate reduction and methanogenesis were inhibited in these tests using sodium molybdate and bromoethanesulfonic acid (BES). The results indicated that iron-reducing bacteria in activated sludge could comprise as much as 3% of the bacteria (*Neilsen et al.* 2002). This finding is interesting as the AS process is an aerobic treatment process, so the reasons for the abundance of these bacteria in this system that can thrive under anaerobic conditions is not well understood. Using fluorescent in situ hybridization in combination with MAR tests showed that all MAR-positive cells hybridized with a *Bacteria*-targeted probe and that 70% did not hybridize with *Proteobacteria*-targeted probes. *Proteobacteria* subclass-targeted probes identified only 20% of these iron reducers as *Gammaproteobacteria* and 10% as *Deltaproteobacteria.* Thus, it appears that we know relatively little about the iron reducing bacteria based on well-studied isolates such as *Shewanella* (*Gammaproteobacteria*) or *Geobacter* (*Deltaproteobacteria*) species.

The effect of the MFC architecture, and in particular the catholyte, on the community that develops in the reactor is not well studied but is likely to be important. The cathode is permeable to oxygen, and the membrane (if present) separating the anode and cathode chambers is permeable to gases and many soluble organic and inorganic species (see Chapter 4). Thus, when oxygen is used as the oxidizer at the cathode, it will diffuse into the anode chamber. Similarly, ammonia, nitrate, sulfate, and other species can diffuse through the membrane. Bacteria will grow using oxygen and alternate electron acceptors (reducing Coulombic efficiencies) and contributing to the diversity of the communities in ways not directly associated with power generation.

High Coulombic efficiencies (*i.e.,* >50%) measured in some MFCs with pure compounds as substrates indicate that most of the substrate is going to cell respiration with energy generation. The amount of substrate converted by exoelectrogens into biomass is not known (*i.e.,* the cell yield), but the highest Coulombic efficiencies are ~85%, suggesting that ~15% of the substrate goes into biomass (see Chapter 7). Therefore, we must be cautious about interpreting the analysis of the community in an MFC as we are not certain to what extent the bacteria present on the anode will be derived from electrogenic activity, versus growth using soluble electron acceptors or growth via fermentation. In most studies the anode community is sampled, but in future studies this should be compared to that found in solution and on the membrane (if present) or cathode (no membrane) in order to better understand the microbial communities that develop in these systems.

2.4.1. MFCs using oxygen at the cathode

An analysis of the communities that were examined in five different systems is summarized in Table 2.1. This analysis does not reveal any specific trend in dominant members of MFC biofilm communities. We can see that *Alpha-*, *Beta-*, or *Gammaproteobacteria* dominated in four different studies, and in the fifth study the main microorganisms could not be identified (Table 2.1). Kim *et al.* (2004) observed a majority of *Betaproteobacteria* clones in the phylogenetically diverse anode community of a two-chamber MFC inoculated with anaerobic sludge and fed wastewater from a starch processing plant. Phung *et al.* (2004) reported a *Betaproteobacteria*-dominated MFC community with a river sediment inoculum enriched with river water. A reactor inoculated with a river sediment using a low concentration of glucose and glutamate produced an *Alphaproteobacteria*-dominated community (*Phung et al.* 2004). An acetate-fed system examined by Lee *et al.* (2003) that was inoculated with activated sludge exhibited a nearly even distribution among the *Alpha-*, *Gamma-*, and *Deltaproteobacteria*.

In one study of the community in an ethanol-fed two-chamber MFC the majority of cloned 16S rRNA gene sequences (83%) were related to *Betaproteobacteria*, primarily *Dechloromonas*, *Azoarcus*, and *Desulfuromonas*, with the balance belonging to the *Deltaproteobacteria* (Fig. 2.5) (*Kim et al.* 2007c). Only a single cloned 16S rRNA gene

Table 2.1 Communities identified in different studies using oxygen at the cathode

Inoculum	Substrate	Community
River sediment (*Phung et al.* 2004)	Glucose+ glutamic acid	65% = *Alpha*-(mainly *Actinobacteria*), 21% = *Beta*-, 3% = *Gammaproteobacteria*, 8% = Bacteroidetes, 3% = others.
River sediment (*Phung et al.* 2004)	River water	11% = *Alpha* -, 46% = *Beta*-(related to *Leptothrix* spp.), 13% = *Gamma*-, 13% = *Deltaproteobacteria,* 9% = *Bacteroidetes,* 8% = others.
Marine sediment (*Logan et al.* 2005)	Cysteine	*Gammaproteobacteria* (40% *Shewanella affinis* KMM), then *Vibrio* spp. and *Pseudoalteromonas* spp.
Wastewater (*Lee et al.* 2003)	Acetate	24% = *Alpha*-, 7% = *Beta*-, 21% = *Gamma*-, 21% = *Deltaproteobacteria;* 27% = others.
Wastewater (*Kim et al.* 2004; *Methe et al.* 2003)	Starch	36% = unidentified, 25% = *Beta*- and 20% = *Alphaproteobacteria*, and 19% = *Cytophaga* + *Flexibacter* + *Bacterioides.*

sequence was similar to a *Geobacter* sequence, and there were no *Shewanella*-like sequences retrieved.

The range of substrates used in these studies, as well as the different internal resistances and Coulombic efficiencies, makes it difficult at this stage to understand what the effect of these factors are on the subsequent community that develops in the reactor. However, it seems clear at this point in our study that exoelectrogenic activity is not limited to just a few microorganisms, despite claims (*Lovley* 2006) that *Geobacter* spp. emerge as a superior competitors in MFC biofilms. It appears unlikely that *Geobacter* or *Shewanella* are the only model microorganisms for exoelectrogenic activity or that these two iron-reducing bacteria capture the microbial diversity that exists in MFC communities. In the five examples summarized in Table 2.1 the main microorganisms were identified as *Actinobacteria*, *Leptothrix*, and *Shewanella* (where such identification was possible), with no *Geobacter*. The notable lack of Geobacter in these systems may be due to the fact that it is an obligate anaerobe, and it is not clear how much oxygen was introduced into the anode chamber (or what the redox conditions were in the system). Side-by-side comparisons of the effect of MFC architecture (*i.e.,* sediment, non-sediment, oxygen, ferricyanide, and poised-potential MFCs), substrate, and inoculum (sediment, river, and wastewater bacteria) are needed to better understand what governs the community that evolves in MFC systems.

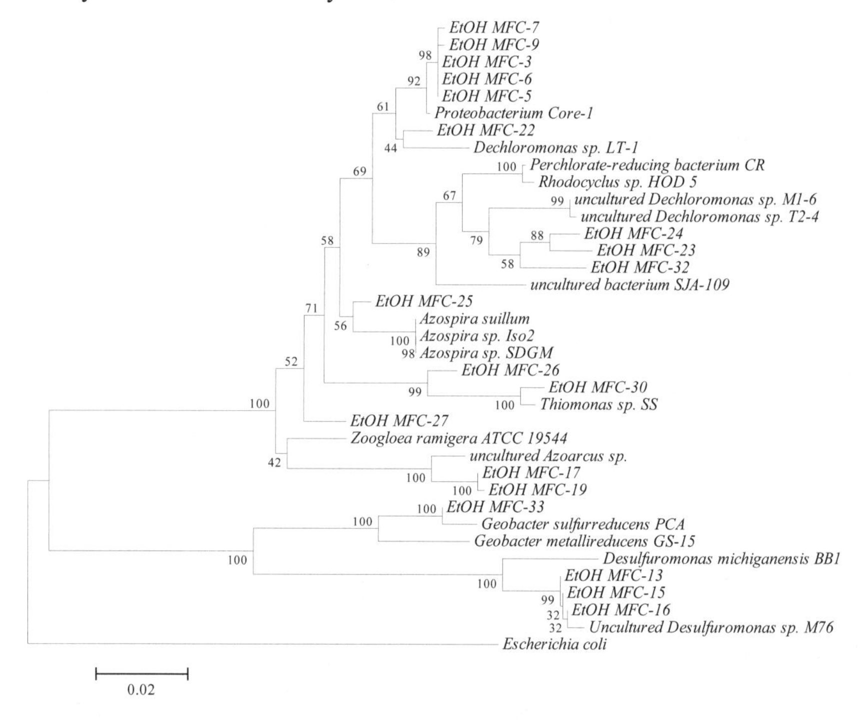

Fig. 2.5 Phylogenetic tree based on a comparative analysis of partial 16S rRNA sequences recovered from an ethanol-fed MFC. [Reprinted from *Kim et al.* (2007c), with permission from Elsevier.]

2.4.2 MFCs with electron acceptors other than oxygen

Using a glucose-fed MFC with ferricyanide at the cathode, Rabaey *et al.* (2004) identified a large number bacterial genera in the biofilm using DGGE (denaturing gradient gel electrophoresis) analysis. The identified sequences were derived from *Firmicutes*, *Gamma-*, *Beta-*, and *Alphaproteobacteria*. Bacteria capable of hydrogen production were predominant, such as the Gram-negative *Alcaligenes faecalis* and Gram-positive *Enerococcus gallinarium*. Conventional plating (nutrient agar) was used to isolate six distinct bacterial species. Based on 16S rRNA sequence analysis, these belong to the *Firmicutes*, *Alpha-*, *Beta-*, and *Gammaproteobacteria*. The electrochemical activity of several isolates was shown to be due to excreted redox mediators, with one of the mediators was identified as pyocyanin, as produced by *Pseudomonas aeruginosa*. Two strains of *P. aeruginosa* isolated from the reactor, and tested in an MFC, produced substantially less power than the mixed culture.

Aelterman *et al.* (2006) examined the microbial community that developed over time in a six-stack MFC fed glucose and using ferricyanide at the cathode. The initial community derived from an anaerobic sludge contained mostly *Proteobacteria*, as well as members of the *Firmicutes* and *Acinobacteria*. Over time, there was a shift in community structure, and analysis of a 16S rRNA gene clone library showed that all clones were similar (>99%) to *Brevibacillus agri*, from the *Firmicutes*. The shift in community was also accompanied by a reduction in internal resistance, suggesting that the dominance of this microorganism increased power due to a reduction in the anode overpotential.

2.4.3 Sediment MFCs

Although oxygen is present at the cathode in a sediment MFC (SMFC), this system differs from other MFCs due to the essentially complete isolation of the bacteria on the anode from oxygen. It is possible in laboratory reactors containing membranes that oxygen can leak across the membrane and thus introduce oxic conditions (even if only briefly). Most likely this occurs at the end of a fed batch cycle, when the substrate has been depleted (although redox conditions within MFCs have not been well investigated). In an SMFC the anaerobic sediment, or mud, stays anaerobic throughout the study, thus ensuring that obligate bacteria will be kept separate from oxygen.

The completely anaerobic conditions of SMFCs may therefore explain why *Deltaproteobacteria* have been observed to dominate the microbial community in most systems. In the first SMFC examined in the laboratory by Bond *et al.* (2002), 71.3 ± 9.6 % of sequences obtained in a 16S rRNA gene clone library were *Deltaproteobacteria*. Most (70%) of these sequences were from a single cluster in the family *Geobacteraceae*, and the microorganism in pure culture most closely related to the sequences obtained from the anode was *Desulfuromonas acetoxidans*. This microbe grows by oxidizing acetate and reducing sulfur. Scrapings of the anode from a SMFC operated at a site in Tuckeron, New Jersey, similarly showed that most bacteria (76%) were *Deltaproteobacteria*, with 59% in the family *Geobacteraceae*, and had >95% 16S rRNA sequence identity to *D. acetoxidans*.

Further examination of the bacteria that develop in SMFCs was made by Holmes *et al.* (2004b) based on a comparison of communities in several marine, salt marsh, and freshwater sediments. Five laboratory and field tests revealed once again that *Deltaproteobacteria* dominated in SMFC reactors (54–76% of gene sequences recovered from the anode). Other predominant sequences originated from *Gammaproteobacteria* (3 cases, 9–10%), *Cytophagales* (33%), and *Firmicutes* (11.6%) (*Holmes et al.* 2004b). It is not the inoculum alone, however, which results in the dominance of the δ-Proteobacteria. Using an air-cathode MFC (*i.e.*, a reactor type that is different than a sediment MFC) inoculated with river and marine sediments produced different results from those observed in these SMFC studies (*Logan et al.* 2005; *Phung et al.* 2004). More likely this community results from selection of particular organisms during the degradation of the complex organic matter and sulfides in the sediment. In contrast to SMFCs where the electron donors are essentially undefined, MFCs tested in the laboratory usually use single substrates.

The bacterial communities that thrive on the anodes of SMFCs also reflect more than just organic matter decomposition, as at least in marine sediments the SMFC environment is usually dominated by sulfide oxidation. In a recent study of the microbial community that developed along an anode near an ocean cold seep, Reimers *et al.* (2006) found that microorganisms closely related to *D. acetoxidans* (*Deltaproteobacteria*) dominated at the 20 to 29-cm depth of the anode (90% of 346 clones), consistent with previous findings of SMFCs. They attributed this community development to the availability of Fe(III) in the surrounding sediments. However, when they examined the communities obtained from more deeply buried sections at 46 to 55-cm depths, they found more diverse populations, with comparable populations of *Epsilonproteobacteria*, *Desulfocapsa,* and *Syntrophus* (23%, 19%, and 16% of clones). At a depth of 70–76 cm, clones were most similar to *Epsilonproteobacteria* and *Syntrophus* (*Deltaproteobacteria*) (32% and 24%). The sulfate-reducing *Deculfocapsa* derive energy for growth from disproportionation of S^0. There was evidence of elemental sulfur on the anode, which they believed was due to sulfide oxidation resulting in electrocatalytic deposition of elemental sulfur, resulting in passivation of the electrode. This deposition was consistent with a decrease in performance of this system over time. As SMFCs are examined in more detail with the surrounding pore water chemistry, we can expect to see a wider range of communities that reflect different organic and inorganic substrates.

2.4.4 Thermophilic MFCs

Few researchers have examined electricity generation outside the normal laboratory temperature range (room temperature to 36°C). Choi *et al.* (2004) examined power production by two thermophilic bacteria (*Bacillus licheniformis* and *B. thermoglucosidasius*) in an MFC. However, power generation required the use of mediators, and thus the bacteria used in this study cannot be considered to be exoelectrogens. One problem with using oxygen at elevated temperatures is that it becomes less soluble as temperature increases. For example, oxygen solubility is 9.0 mg/L at 20°C, decreasing to 7.5 mg/L at 30°C and 6.0 at 50°C. Jong *et al.* (2006) overcame this potential oxidant limitation by recirculating oxygen saturated water to the cathode, achieving 1.03 W/m^2 at 55°C. Using 16S rRNA gene analysis, they found that one bacterium (57.8% of 199 total clones) was phylogenetically related to an uncultured clone, E4 (GenBank accession number AY526503, 99% homology) and distantly related to *Deferribacter desulfuricans*. This microbe is a sulfur-, nitrate-, and arsenate-reducing thermophile isolated from a deep-sea hydrothermal vent. The next most dominant pattern (15.1%) was related to a *Coprothermobacter* spp. There was a complete absence of sequences from *Proteobacteria* in their clone libraries. Of the various studies conducted

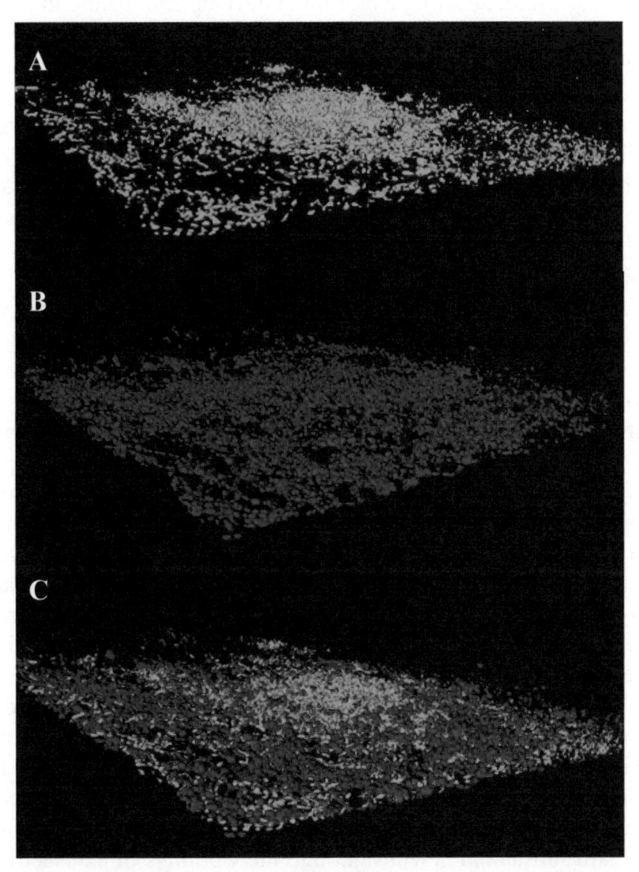

Fig. 2.6 Composite images of an electrochemically active mixed-culture biofilm growing on a carbon electrode obtained using confocal scanning laser microscopy, with live (green) and dead (red) stained cells. (A) The live-only image shows a mound of bacteria that are alive, even when distant from the surface; (B) dead cells are uniformly distributed as many bacteria in the inoculum that may have adhered to the surface could not use it as an electron acceptor; (C) the combined image. [From *Logan and Regan* (2006a), reprinted with permission from Elsevier.]

on communities in MFCs, this study by Jong *et al.* showed the least diversity suggesting that there is a relatively less diverse community capable of functioning in MFCs at elevated temperatures. However, substantially more work in this area would need to be done to support that speculation.

2.5 MFCs as tools for studying exoelectrogens

While power generation is the main goal of MFC studies, the system affords the scientist an interesting and novel platform for examining the microbial ecology of exoelectrogenic bacteria. When bacteria degrade an insoluble metal, the features of the surface change over time, and the water chemistry becomes more complex as the relative oxidized/reduced concentration ratios of the metal change over time. In an MFC, the electrode is non-corrosive, allowing a biofilm to develop and mature in a manner that can be more easily investigated, especially using microscopy tools. The composition of a bacterial community can be investigated using fluorescent in situ hybridization (FISH) probes coupled with confocal scanning laser microscopy (CSLM). Little work has been done in this area with MFCs so far, but we can expect such studies to emerge soon.

The viability profile of mature mixed-culture MFC biofilms was recently inspected using BacLight viability staining and visualization with CSLM (Fig. 2.6). In general, cells indicated to be non-viable by this approach (red) were uniformly distributed across the surface. However, live (green) cells appeared in many locations to exist as a protruding colony, suggesting that cells were viable at relatively large distances (*i.e.,* many layers of microorganisms) from the electrode. If these bacteria were active, it becomes an interesting question to probe about how far bacterial growth can be sustained—via nanowires or mediators?—from the electrode surface. There are possible alternative explanations for the observation of viable microbes distant from the surface such as the presence of fermentative bacteria metabolizing a substrate without the need for the anode.

Microbial ecology of MFC biofilms. One of the most significant observations about the stability of an MFC is that the anode does not appear to foul over time. That is, the bacteria at the surface must remain viable and able to continually use the surface. Thus, it is reasonable to hypothesize that there are mechanisms by which bacteria can invade a surface and, by examples in other ecological systems, use different colonization strategies and occupy unique niches leading to a succession of bacterial species with depth in the biofilm. These niches will be in part related to proximity of the surface to the bacteria. The analysis of these amazingly complex and yet poorly described exoelectrogenic biofilm communities can best be examined using MFCs, as these devices allow easy growth of the bacteria for study, and they provide surfaces most amenable to direct visual analysis as the anode surface is flat and non-corrosive (graphite sheet), permitting examination using FISH and confocal microscopy. It is expected that in the next few years, we will gain substantial new insight into the workings of exoelectrogenic biofilms and the microbial ecology of different bacteria in these systems using MFCs as a platform for these ecological studies.

CHAPTER 3

Voltage Generation

3.1 Voltage and current

The first observation you make about any electrical source is the voltage. Appliances operate on 110 V in the US, while in Europe 220 V is standard. Batteries are commonly 6 V or 9 V depending on the application, while smaller devices and cell phones vary in voltage requirements (my cell phone is 3.6 V). The voltage produced by batteries and fuel cells is often increased by linking them in series. In a flashlight, for example, you often place two or more batteries in series to increase the total voltage, while in a toy batteries can be linked in parallel to increase current. Hydrogen fuel cells operate at a single cell working voltage of ~0.7 V so they are stacked together to produce higher voltages. Voltages can be increased using DC-DC converters, and converted from DC to AC.

Microbial fuel cells commonly achieve a maximum working voltage of 0.3–0.7 V. The voltage is a function of the external resistance (R_{ex}), or load on the circuit, and the current, I. The relationship between these variables is the well-known equation

$$E = I\,R_{ext} \tag{3-1}$$

where we use E for the cell potential (V is also used for voltage, although the symbol V and the units V = Volts can lead to confusion). The current produced from a single MFC is small, so that when a small MFC is constructed in the laboratory the current is not measured, but instead it is calculated from the measured voltage drop across the resistor as $I = E/R_{ex}$. The highest voltage produced in an MFC is the open circuit voltage, OCV, which can be measured with the circuit disconnected (infinite resistance, zero current). As the resistances are decreased, the voltage decreases. The power at any time is calculated as $P = IE$, as described in more detail in the next chapter.

The voltage generated by an MFC is far more complicated to understand or predict than that of a chemical fuel cell. In an MFC it takes time for the bacteria to colonize the electrode and manufacture enzymes or structures needed to transfer electrons outside the cell. In mixed cultures, different bacteria can grow, setting different potentials. As discussed below, the potential even for a pure culture cannot be predicted. However, there are limits to the maximum voltages that can be generated based on thermodynamic relationships for the electron donors (substrates) and acceptors (oxidizers). We therefore review these relationships to obtain a better understanding of the boundaries on MFC operation.

3.2 Maximum voltages based on thermodynamic relationships

The maximum electromotive force, E_{emf}, that can be developed in any type of battery or fuel is given by

$$E_{emf} = E^0 - \frac{RT}{nF} \ln(\Pi) \tag{3-2}$$

where E^0 is the standard cell electromotive force, R = 8.31447 J/mol-K the gas constant, T the absolute temperature (K), n the number of electrons transferred, and F = 96,485 C/mol is Faraday's constant. The reaction quotient is the ratio of the activities of the products divided by the reactants raised to their respective stoichiometric coefficients, or

$$\Pi = \frac{[\text{products}]^p}{[\text{reactants}]^r} \tag{3-3}$$

By the IUPAC convention, all reactions are written in the direction of chemical reduction, so that the products are always the reduced species, and the reactants are the oxidized species (oxidized species + e- → reduced species). Also by IUPAC convention, we take as standard conditions a temperature of 298 K, and chemical concentrations of 1 M for liquids and 1 bar for gases (1 bar = 0.9869 atm = 100 kPa). All values of E^0 are calculated with respect to that of hydrogen under standard conditions, which is defined to be $E^0(H_2)$ = 0, referred to as the "normal hydrogen electrode" (NHE). Thus, the standard potentials for all chemicals is obtained with $\Pi = 1$ relative to a hydrogen electrode.

In biological systems the reported potentials are usually pre-adjusted to neutral pH, because the cytoplasm of most cells is at pH=7. For hydrogen, with $2H^+ + 2e^- \rightarrow H_2$, this means that the adjusted potential at 298 K is

$$E^{0/} = E^0 - \frac{RT}{nF} \ln \frac{H_2}{[H^+]^2} = 0 - \frac{(8.31\,\text{J/mol K})\,(298.15\,\text{K})}{(2)\,(9.65\times 10^4\,\text{C/mol})} \ln \frac{[1\,\text{bar}]}{[10^{-7}\,\text{M}]^2} = -0.414\ \text{V} \tag{3-4}$$

where the $^/$ on E is used to denote the pH-adjusted "standard" condition commonly used by microbiologists. In most calculations, therefore, the hydrogen potential is not zero as a result of the assumption of all species being present in a pH = 7 solution. These potentials need to be adjusted for other temperatures or pressures, or pH if different from 7. At 303 K (30°C), a temperature commonly used in the laboratory for incubation of bacteria, for example, $E^{0/}(H_2) = -0.421$ V.

For hydrogen (H^+/H_2), chemicals that will be oxidized by H^+ have more negative potentials, while those that are reduced by H_2 have more positive potentials. For example,

H_2 is oxidized by oxygen. The half reaction for oxygen is $\frac{1}{2}O_2 + 2H^+ + 2e^- \rightarrow H_2O$ and $E^0\ (O_2) = 1.229$ V, so the adjusted value for oxygen at pH = 7 is

$$E^{0/} = E_0 - \frac{RT}{nF} \ln \frac{1}{[O_2]^{1/2}[H^+]^2} \tag{3-5}$$

$$E^{0/} = 1.229 - \frac{(8.31\,\text{J/mol K})\,(298.15\,\text{K})}{(2)\,(9.65 \times 10^4\,\text{C/mol})} \ln \frac{1}{[0.2]^{1/2}[10^{-7}\,\text{M}]^2} = \quad 0.805\ \text{V} \tag{3-6}$$

Table 3.1 Anode and cathode potentials for different anodic and cathodic reactions. $E^{/0}$ values are adjusted for pH = 7 at 298 K except as indicated

Electrode	Reaction	E^0 (V)	Conditions	$E^{/}$ (V)
Anode				
A-1	$2\ H^+ + 2\ e^- \rightarrow H_2$	0.000	pH = 7	−0.414
A-2	$2HCO_3^- + 9H^+ + 8e^- \rightarrow CH_3COO^- + 4H_2O$	0.187	HCO_3^- = 5 mM, CH_3COO^- = 16.9 mM, pH = 7	−0.300
A-3 (*Logan et al.* 2006)			HCO_3^- = 5 mM, CH_3COO^- = 5 mM, pH = 7	−0.296
A-4 (*Rittmann and McCarty* 2001)	$CO_2 + HCO_3^- + 8H^+ + 8e^- \rightarrow CH_3COO^- + 3H_2O$	0.130	pH = 7	−0.284
A-5 (*Rittmann and McCarty* 2001)	$6CO_2 + 24H^+ + 24e^- \rightarrow C_6H_{12}O_6 + 6H_2O$	- 0.014	pH = 7	−0.428
Cathode				
C-1 (*Logan et al.* 2006)	$O_2 + 4H^+ + 4\ e^- \rightarrow 2H_2O$	1.229	pO_2 = 0.2, pH = 7	0.805
C-2 (*Logan et al.* 2006)			pO_2 = 0.2, pH = 10	0.627
C-3 (*Logan et al.* 2006)	$O_2 + 2H^+ + 2e^- \rightarrow H_2O_2$	0.695	pO_2 = 0.2, H_2O_2 = 5 mM, pH = 7	0.328
C-4			pO_2 = 0.2, H_2O_2 = 0.22 mM, pH = 7	0.370
C-5 (*Logan et al.* 2006)	$Fe(CN)_6^{3-} + e^- \rightarrow Fe(CN)_6^{4-}$	0.361	$Fe(CN)_6^{3-}$ = $Fe(CN)_6^{4-}$	0.361
C-6 (*Logan et al.* 2006)	$MnO_2(s) + 4H^+ + 2e^- \rightarrow Mn^{2+} + 2H_2O$	1.229	Mn^{2+} = 5 mM, pH = 7	0.470
C-7 (*You et al.* 2006)	$MnO_4^- + 4H^+ + 3e^- \rightarrow MnO_2 + 2H_2O$	1.70	MnO_4^- = 10 mM, pH = 3.5	1.385
C-8 (*ter Heijne et al.* 2006)	Fe^{3+} e- $\rightarrow Fe^{2+}$ (low pH)	0.77	$Fe^{3+} = Fe^{2+}$, T = 303 K (low pH)	0.78
C-9 (*Madigan and Martinko* 2006)			T = 303 K (neutral pH)	0.20

The activity of a pure liquid or a solid is constant, so here the activity of water is unity. Because $E^{0/}(O_2) > E^{0/}(H_2)$, oxygen is reduced by hydrogen. When the voltage is positive, the reaction is exothermic. The calculations for can also be expressed in terms of the change in Gibbs free energy, ΔG_r^0 [J], as

$$E^0 = -\frac{\Delta G_r^0}{nF} \tag{3-7}$$

Note here that the reaction is exothermic when ΔG_r^0 is negative. A more detailed analysis of thermodynamic calculations is available in standard physical chemistry textbooks and specialized texts on biochemical reactions (*Alberty* 2003; *Rittmann and McCarty* 2001).

Total cell potential. The total potential that can be produced by any fuel cell is the difference in the anode and cathode potentials, or $E_{emf} = E_{Cat} - E_{An}$. For the adjusted standard conditions of pH = 7, this is

$$E_{emf}^{/} = E_{Cat}^{0/} - E_{An}^{0/} \tag{3-8}$$

For example, let us assume that oxygen is used at the cathode, and hydrogen at the anode. For this case at the given conditions (298 K, 1 bar, pH = 7), this is $E^{0/}_{emf}$ = 0.805 V − (−0.414 V) = 1.219 V. In a hydrogen fuel cell, we might have conditions of 80°C and 2 bar for oxygen and hydrogen, and a pH in the cation exchange membrane (CEM) of 3, so this would produce a slightly higher potential of E_{emf} = 1.24 V. More specific examples that apply to MFCs are discussed below.

Potentials other than NHE. While it is useful to express all potentials relative to a normal hydrogen electrode (NHE), most experiments are conducted using Ag/AgCl or calomel reference electrodes. For converting voltages obtained with a Ag/AgCl electrode to NHE, it depends on the specific solution in the probe, but typically to get NHE add 0.195 V (*Liu and Logan* 2004) or 0.205 V (*ter Heijne et al.* 2006). For converting the voltage obtained from a standard calomel electrode (SCE), add 0.241 V to get NHE.

3.2.1 Anode

If thermodynamics limits overall power production, we can expect that the measured anode potential will approach that of the calculated maximum potential (*i.e.,* the potential set by substrate oxidation). As noted above, the maximum voltage is produced in open-circuit mode, so the maximum potential should be close to that of the open-circuit potential (*OCP*). Most MFCs operating on a variety of substrates produce an *OCP* approaching −0.3 V (vs. NHE). For acetate, we have the HCO_3^-/Ac couple expressed as a reduction as:

$$2\,HCO_3^- + 9\,H^+ + 8e^- \rightarrow CH_3COO^- + 4\,H_2O \tag{3-9}$$

For acetate $E^0 = 0.187$ V, with a concentration of 1 g/L (16.9 mM) and under conditions of neutral pH = 7 and an alkalinity set by the bicarbonate concentration of $HCO_3^- = 5$ mM, we have

$$E_{An} = E^0_{An} - \frac{RT}{8F} \ln \frac{[CH_3COO^-]}{[HCO_3^-]^2\,[H^+]^9} \tag{3-10}$$

$$E_{An} = 0.187 - \frac{(8.31\,\text{J/mol K})\,(298.15\,\text{K})}{(8)\,(9.65\times10^4\,\text{C/mol})} \ln \frac{[0.0169]}{[0.005]^2[10^{-7}\,\text{M}]^9} = -0.300\text{ V} \tag{3-11}$$

Slightly different results are obtained if CO_2 is included in the reaction for acetate, and a slightly lower (more negative) potential is predicted with glucose (Table 3.1). For both cases we see an anode potential predicted by these calculations is very close to the measured anode *OCP* with acetate. This suggests that the conditions inside the cell somehow change to achieve this potential, as further explored in the next section.

Example 3.1

As substrate is consumed, protons are produced at the anode and the pH can drop. If we assume that the system is well buffered, the pH will not change. For acetate, calculate the potentials at the anode (a) if pH does not change, and (b) if pH changes in proportion to the consumption of substrate for an initial acetate concentration of 1 g/L.

(a) For the first case, we note from stoichiometry that two moles of bicarbonate will be produced per mole of acetate. Thus, we repeat the above calculation at several different acetate concentrations always keeping the pH constant. The results are shown in the graph below, where it can be seen that at 90% conversion, $E^0_{An} = -280$ V, and further if only 1 nM of acetate remains (last point on *x*-axis), $E^0_{An} = -0.233$. Thus, the range of change of acetate in terms of potential is only 22% over this range.

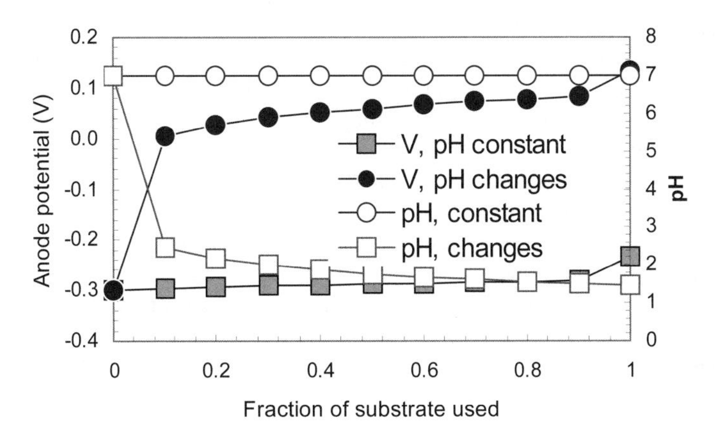

(b) If the pH is allowed to change over the same cycle as calculated above, then the pH quickly drops resulting in a very rapid increase of the anode potential to E^{0}_{An} = 0.005 V with only 10% of the acetate consumed, with E^{0}_{An} = 0.135 V when 1 nM of acetate remains.

These results point to a very important point that buffering of pH change and removal of the produced protons is very important for maintaining the potential of the anode that will help to produce the greatest voltage in the MFC.

3.2.2 Cathode

For an MFC using oxygen, the cathode potential is a maximum of $E^{0/}_{Cat}$ = 0.805 V. Thus, for an air-cathode MFC with 1 g/L of acetate (16.9 mM) as substrate (HCO_3^- = 5 mM, pH = 7), the maximum cell potential is $E^{/}_{cell}$ = 0.805 V − (−0.300 V) = 1.105 V. However, the cathode potential with oxygen is much less in practice than predicted here. Typically, the *OCP* of an air cathode is ~0.4 V, with a working potential of ~0.25 V even with a Pt catalyst. In one set of tests (Fig. 3.1) the OCP_{Cat} of an MFC lacking a CEM was 0.425 V (0.230 V vs. Ag/AgCl) (*Liu and Logan* 2004). Hot pressing a CEM (Nafion™) to the cathode substantially reduced the cathode potential to OCP_{Cat} = 0.226 V. The anode *OCP* was −0.275 V, with a working anode potential of *ca.* −0.205 V (−0.400 V vs. Ag/AgCl), either in the presence or absence of the CEM.

The reduction of oxygen to water requires a four-electron transfer, but that may not always be achieved. It is also possible that oxygen reduction results in hydrogen peroxide production, and this is only a two-electron transfer reaction. The standard potential for H_2O_2 evolution is 0.695 V, but under conditions reasonable for MFCs we obtain a result of $E^{0/}$ = 0.328 V (Table 3.1). This is much closer to the *OCP* observed for the cathode of 0.425 V. Production of H_2O_2 is problematic as it is a strong oxidizer, and can result in degradation of the electrode or a membrane. However, its production might have a useful result as it could serve function as a disinfectant for limiting biofilm formation on the cathode or perhaps in helping to degrade sorbed organics. The form of the oxygen produced on the MFC cathode, and its role on the long-term performance of the cathode needs to be better explored.

All other potential oxidizers listed in Table 3.1 are non-sustainable as they must be chemically regenerated. The most commonly used chemical catholyte in MFC fuel cells next to oxygen is ferricyanide, or hexacyanoferrate, $Fe(CN)_6^{3-}$. It has a standard potential of 0.361 V, is highly soluble in water, and it does not require a precious metal on the cathode such as Pt. Tests using ferricyanide show much greater power generation than those with oxygen due to the fact that there is little polarization of the cathode so that the cathode potential achieved is quite close to that calculated for standard conditions (*ter Heijne et al.* 2006; *You et al.* 2006). Thus, while oxygen is predicted to have a higher cathode potential than ferricyanide, in practice the potentials achieved using oxygen are much lower than theoretical values. In two-chamber MFC tests Oh and Logan (2006) found that replacing the aqueous cathode using oxygen with ferricyanide increased power by 1.5 to 1.8 times, although power densities produced in this system were low due to the high internal resistance of the system. Rabaey *et al.* (2004) achieved one of the highest power densities yet produced in an MFC using a ferricyanide catholyte (4.1 W/m^2 of anode surface area), but they did not report on power production in that system with dissolved oxygen.

Other catholytes that have been used are iron, manganese, and permanganate. Ter Heijne *et al.* (2006) devised a system using a bipolar membrane with ferric iron converted to ferrous iron at the cathode (Table 3.1), producing a maximum power density of 0.86 W/m^2-anode. They reported a cathode *OCP* of 0.674 V (vs. NHE) using ferric iron sulfate but did not measure the relative concentrations of Fe^{3+}/Fe^{2+}. Assuming an equal concentration of these species, the maximum cathode potential would be E_{cat} = 0.780 V. This is a difference of only 0.106 V, which is quite good in comparison to oxygen where the difference between the cathode *OCP* and E_{cat} is much larger. Also, this cathode potential for iron is higher than that typically achieved with oxygen.

Shantaram *et al.* (2005) developed a fuel cell based on a sacrificial metal anode (Mg) and a cathode made of solid MnO_2. Although this is not a true MFC in the manner described here (no bacteria on the anode), the reduced Mn^{2+} produced at the cathode was assumed to be biologically regenerated due to bacterial oxidation of the released manganese. You *et al.* (2006) used permanganate, which has E^0_{Cat} = 1.70 V, as a catholyte. Under the conditions tested, the potential of the cathode (Table 3.1) could be $E^{0/}_{Cat}$ = 1.385 V (total circuit potential of E_{emf} = 1.53 V). The cathode *OCP* was 1.284 V, one of the highest reported for an MFC system. Based on these results, only 0.101 V was lost at the cathode. Their highest power density was 4.0 W/m^2 of anode surface area. In contrast, the same reactor with ferricyanide as the catholyte achieved 1.2 W/m^2. This comparison illustrates the important role of the *OCP* on power generation, as discussed more extensively in the next chapter.

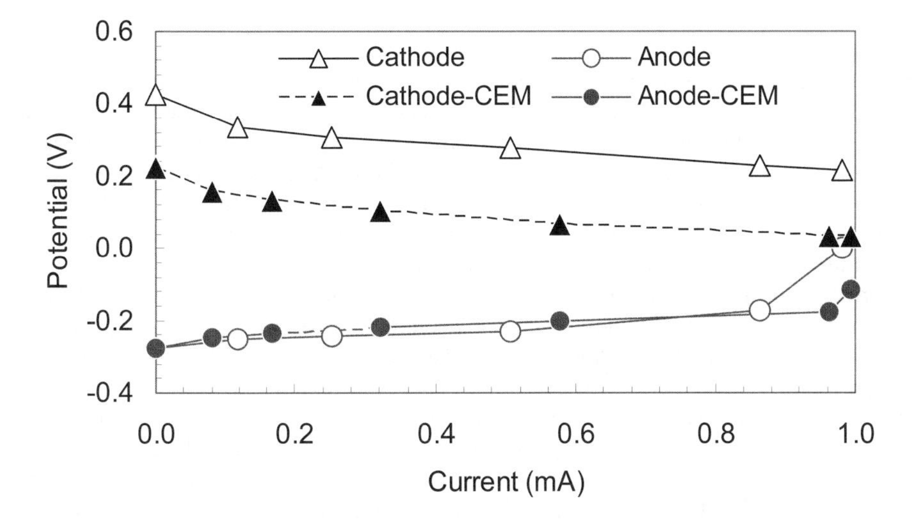

Fig. 3.1 Anode and cathode potentials (vs. NHE) in a glucose-fed MFC in the presence and absence of a CEM Note that the cathode potentials were substantially reduced due to the presence of the CEM. [Reprinted from *Liu and Logan* (2004), with permission from the American Chemical Society.]

Table 3.2 Potentials of different couples (pH = 7, except as noted; temperature is assumed to be 303 K based on value reported for H_2 couple) (*Madigan and Martinko* 2006)

Couple	Potential (V)	Couple	Potential (V)
CO_2/glucose, 24 e^-	−0.43	$S_4O_6^{2-}/S_2O_3^{2-}$, 2 e^-	0.024
$2H^+/H_2$, 2 e^-	−0.42	Fumarate, succinate, 2 e^-	0.030
CO_2/methanol, 6 e^-	−0.38	Cytochrome $b_{ox/red}$, 1 e^-	0.035
NAD^+/NADH, 2 e^-	−0.32	Ubiquinone$_{ox/red}$, 2 e^-	0.11
CO_2/acetate, 8 e^-	−0.28	Fe^{3+}/Fe^{2+}, 1 e^- (pH = 7)	0.20
S^0/H_2S, 2 e^-	−0.28	Cytochrome $c_{ox/red}$, 1 e^-	0.25
SO_4^{2-}/H_2S, 8 e^-	−0.22	Cytochrome $a_{ox/red}$, 1 e^-	0.39
Pyruvate/lactate, 2 e^-	−0.19	NO_3^-/NO_2^-, 2 e^-	0.42
		$NO_3^-/½N_2$, 5 e^-	0.74
		Fe^{3+}/Fe^{2+}, 1 e^- (pH = 2)	0.76
		½ O_2/H_2O, 2 e^-	0.82

3.3 Anode potentials and enzyme potentials

While the potentials of the anode and cathode set the limits for the maximum voltage achievable for power generation, using the potentials for these substrates does not fully consider the biochemical basis for power generation. Bacteria that use oxygen and many alternate electron acceptors such as iron use the citric acid cycle for oxidation of substrates, which results in the oxidation of the substrate with the production of three different electron carriers (NADH, FADH, and GTP) (*Madigan and Martinko* 2006). The ATP yield under aerobic conditions is the highest achievable due to the largest potential between NADH and oxygen. For example, under aerobic conditions when glucose is first oxidized to pyruvate (producing a net of 2 ATP), each pyruvate is decarboxylated to acetyl-CoA (producing 1 NADH = 3 ATP), and then the Citric Acid Cycle (CAC) is used to completely oxidize the pyruvate producing 1 GTP (1 ATP), 4 NADH (12 ATP), and 1 FADH (2 ATP). With oxygen, the 15 ATP from each pyruvate in the CAC (total of 30 ATP), plus $2 \times 3 = 6$ ATP from pyruvate oxidation and 2 from glycolysis nets a maximum of 38 ATP under aerobic conditions (*Madigan and Martinko* 2006). ATP is generated due to the pumping of protons across the inner cell membrane by the respiratory enzymes. When they flow back through ATPase, ATP is generated from ADP.

When oxygen is used as the terminal electron acceptor, a total of five protons are pumped across the inner membrane for *Paracoccus denitrificans* or *Escherichia coli* (although some different respiratory enzymes are involved for these two microorganisms). When nitrate is used as a terminal electron acceptor by *E. coli*, only four protons are pumped across the inner membrane, and thus the yield of ATP will be reduced with nitrate compared to that with oxygen. This is because there is less energy available with nitrate than oxygen, as indicated by a lower redox potential for nitrate ($NO_3^-/½N_2$, $E^{0/} =$ 0.74 V) than oxygen ($½O_2/H_2O$, $E^{0/} = 0.82$ V under the same conditions) (Table 3.2). Less energy means fewer protons can be pumped across the membrane.

The important feature of this biochemical pathway analysis is that for energy generation using the CAC it is actually NADH which is the electron donor into the respiratory chain, not acetate. While the discussion in the previous section has therefore focused on energy transfer from the substrate (acetate) to oxygen, we can see that the

NADH/NAD$^+$ couple is the chemical species directly involved with the terminal electron acceptor (*i.e.*, the anode under electrogenic conditions in an MFC or oxygen under aerobic conditions).

NAD$^+$ accepts two electrons and two protons, forming NADH. Under standard conditions (adjusted to pH = 7) the redox potential of NADH/NAD$^+$ is $E^{0/} = -0.32$ V (Table 3.2). If we compare that potential to the acetate couple, $E^{0/} = -0.30$ V (under the non-standard conditions shown in Table 3.1), we see that this implies that acetate cannot be oxidized to produce NADH, as the reaction is unfavorable ($E = -0.32\text{ V} - (-0.30\text{ V}) = -0.02$ V). However, as we have seen acetate oxidation coupled to oxygen is favorable. How then does the reaction proceed to successfully produce NADH? The answer is that the relative concentration of NADH and NAD$^+$ can be regulated within the cell to achieve the needed potential.

The NADH/NAD+ reaction can be written as

$$\text{NAD}^+ + 2\text{H}^+ + 2\text{e}^- \rightarrow \text{NADH} \qquad (3\text{-}12)$$

The potential of this half reaction, assuming equal concentrations of NADH and NAD+ (*i.e.*, 1 M concentrations under standard conditions) and also assuming a pH = 7, can be calculated as

$$E^{0/} = E^0 - \frac{RT}{2F}\ln\frac{[\text{NADH}]}{[\text{NAD}^+][\text{H}^+]^2} = E^0 - \frac{RT}{2F}\ln\frac{[1]}{[1][\text{H}^+]^2} \qquad (3\text{-}13)$$

Using this equation with $E^{0/} = -0.32$ V, we can calculate that $E^0 = 0.094$ V at pH = 1. Even if we do not know the true concentrations of NADH or NAD$^+$, we can for an example calculate the potential assuming the ratio of N = NADH/NAD$^+$ = 0.1, and we have at a pH = 7

$$E^{0/} = 0.094\text{ V} - \frac{RT}{2F}\ln\frac{[0.1]}{[1][10^{-7}]^2} = -0.290\text{ V} \qquad (3\text{-}14)$$

Thus we see now that the reaction becomes favorable with acetate as $E = -0.290\text{ V} - (-0.300\text{ V}) = 0.010$ V. We can see using this example how regulation of the relative intracellular pools of NADH and NAD+ allow the cell to achieve virtually any potential—as long as it is below that of the starting substrate. In our example, if the NADH pool builds up to be too large, then the CAC cycle must cease as it will ultimately become an unfavorable (endothermic) reaction. This will stop cell respiration until the NADH pool can be "relieved" through conversion back to NAD$^+$. We can anticipate that this is what happens when an MFC is put into open circuit mode. The concentrations of reduced species, such as NADH, build up within the bacteria until such point as the potentials become unfavorable and respiration stops. Once we connect the MFC circuit, the "pressure" is released; *i.e.,* the reduced species can be oxidized and the electrons can flow to the terminal electron acceptor (the anode, and then on to the cathode).

While this above situation is illustrative, it is by no means certain in terms of critical species being NADH/NAD$^+$! There are in fact many different oxidized and reduced species in the respiratory chain of bacteria, and at this point our understanding of where

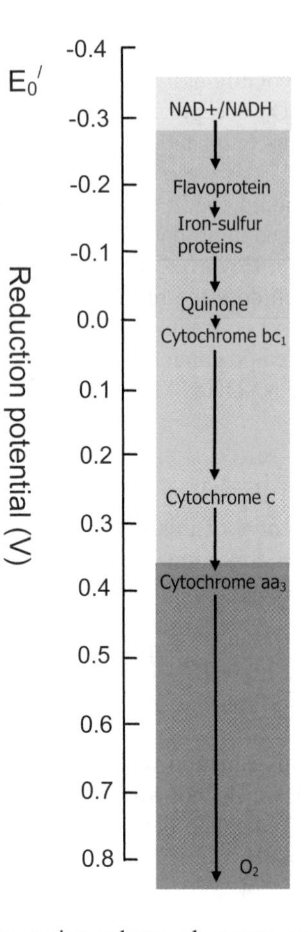

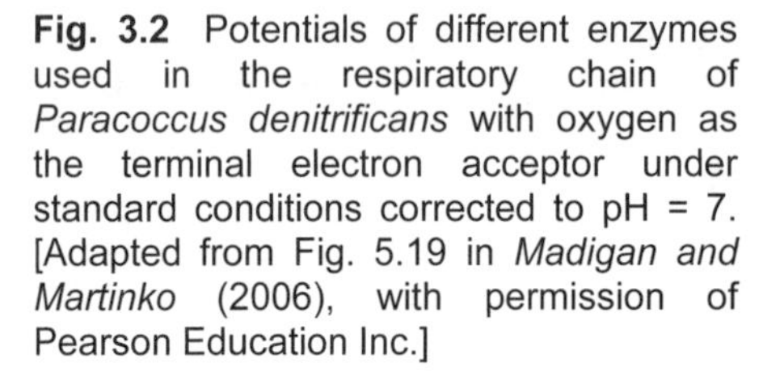
Fig. 3.2 Potentials of different enzymes used in the respiratory chain of *Paracoccus denitrificans* with oxygen as the terminal electron acceptor under standard conditions corrected to pH = 7. [Adapted from Fig. 5.19 in *Madigan and Martinko* (2006), with permission of Pearson Education Inc.]

reduced species holding electrons build up—or the point where electrons exit the respiratory chain—is only speculative. The series of respiratory enzymes used by *P. denitrificans*, along with their potentials at pH = 7, is shown in Fig. 3.2. This pathway shows the presence of three cytochromes (bc_1, *c,* and aa_3). Other bacteria can use these or other cytochromes as well as other respiratory enzymes, and thus this pathway varies among bacteria. Also, many bacteria have branched respiratory pathways allowing them to use different cytochromes for different electron acceptors (*Madigan and Martinko* 2006). For example, *Shewanella* oneidensis has 39 *c*-type cytochromes, which was more than any other microorganism sequenced in 2002 (*Pseudomonas aeruginosa*, 37; *Vibrio cholerae*, 12; *E. coli*, 7) (*Heidelberg et al.* 2002). There were 14 *c*-type cytocromes identified with four or more heme-binding sites that had not been previously described for this strain. *Geobacter sulfurreducens* was subsequently sequenced, and it was found to contain 73 c-type cytochromes that contained two or more heme groups (*Methe et al.* 2003).

Example 3.2

It was shown above that the N = NADH/NAD$^+$ ratio will produce different potentials for this half cell reaction. Develop a graph that shows the extent of the change in the potential of the ratio of these two species over a range of 10^{-3} to 10^3 at 298 K for a pH of 7 and 5.

To do this, we recast the above equation in terms of this ratio, b_N, producing the equation at a pH = 7 of: $E^{0/}$

$$E^{0/} = 0.094\,\text{V} - \frac{RT}{2F}\ln\frac{b_N}{[10^{-7}]^2} = 0.094\,\text{V} - \frac{(8.31\,\text{J/mol K})(298.15\,\text{K})}{(2)(9.65\times10^4\,\text{C/mol})}\ln\frac{b_N}{[10^{-7}]^2}$$

$$E^{0/} = 0.094\,\text{V} - 0.01285\ \ln(10^{14} b_N)$$

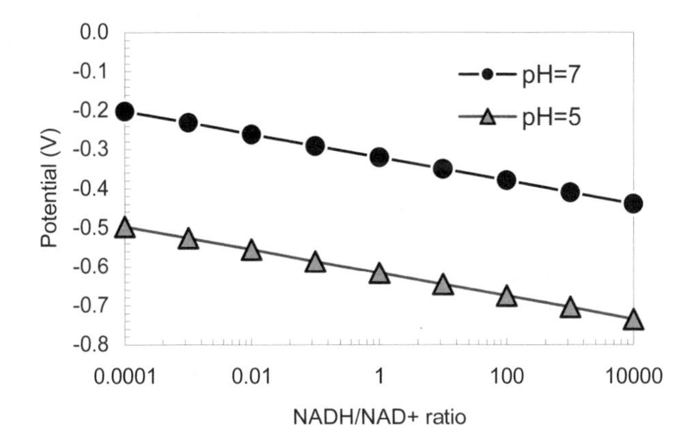

As shown in the figure above, the potential ranges from −0.202 V to −0.438 V for pH = 7. Thus, we can see that there is a large range of potentials possible depending on the oxidation state of the NADH at the intracellular pH = 7 expected for most cells.

Repeating this calculation for a pH = 5, we see that the potentials are lowered (more negative). This is not desirable, however, as this means the potentials are too low for acetate to be used as substrate (E_{An}= −0.166 V for the conditions given in eq. 3-11 at pH = 5). Therefore we can infer a reason why near-neutral intracellular pH values are useful for metabolism when NAD is the electron carrier.

NADH levels in the cell are central to the regulation of many different processes, and thus it may not possible for bacteria to vary these ratios outside of certain ranges. Leonardo *et al.* (1996) measured NADH/NAD$^+$ ratios of 0.094 for *E. coli* using glucose under aerobic conditions, but under anaerobic conditions the ratio was 0.22. For *Clostridium kluyveri*, it has been found that under steady conditions the ratio is 0.27 and that the magnitude of the ratio NADH/NAD$^+$ affects the shift to solventogenesis (*Thauer et al.* 1977). In *C. acetobutylicum* the ratio was noted to vary from 0.1 to 0.79 in 5 different studies, with a ratio of 1.0 resulting in complete inhibition of glyceraldehyde-3-

phosphate dehydrogenase (GAPDH) which can shut down glucose or glycerol fermentation (conversion of Glyceraldehyde-3-P to pyruvate) (*Girbal et al.* 1995). So far, there has been no investigation of these ratios in known exoelectrogenic strains of bacteria.

3.4 Role of communities versus enzymes in setting anode potentials

The above discussion of potentials determined by the relative abundance of the NAD^+ in the oxidized and reduced forms gives us some insight into the factors that can affect the microbial ecology in an MFC. First, let us consider the case of a pure culture. It is certainly likely that the concentration of NAD^+, as well as the relative ratio of other oxidized and reduced species, are regulated by the cell. One microorganism might only be able to achieve a certain $NADH/NAD^+$ ratio compared to another, setting a limit on the potential that a specific microbial species can set relative to the circuit. Thus, the anode potential could vary among bacteria, setting different possible scenarios of power generation for different strains of bacteria. As a goal of bacteria is to obtain the most energy possible from degradation of a substrate, it is easy to envision that a high $NADH/NAD^+$ ratio benefits the microbe. By similar reasoning, a low ratio for the concentration of reduced to oxidized species is desirable for the terminal enzyme that transfers the electrons to the anode electrode. In this manner, the largest possible potential is set, resulting in the largest potential energy gain for the cell.

Let us consider what happens in a co-culture where two different bacteria (species 1 and 2) compete to use the electrode. Both could try to maximize the $NADH/NAD^+$ ratio in order to gain the advantage of a large potential, so let us say that this value is −0.261 V. Now let us assume that species 1 has regulated its terminal enzyme so that a final potential of 0.20 V is produced (this is the potential for the Fe^{3+}/Fe^{2+} couple standard conditions at pH = 7). Thus, the potential for species 1 is $E = 0.20\text{ V} - (-0.26\text{V}) = 0.46\text{ V}$. This leaves a potential for us to harvest the remaining potential of $E = 0.80\text{ V} - (0.20\text{ V}) = 0.6\text{ V}$ in an MFC (if we could obtain that potential using oxygen). However, let us say that species 2 now sets a less-negative potential of −0.10 V for the anode. Species 2 will now get less potential from the pathway ($E = 0.16$ V) but the important result is that species 1 can no longer transfer electrons from its enzyme, which is at 0.20 V to the electrode as that is at an overall negative potential (*i.e.,* −0.10 V). Thus, species 1 must adjust the relative oxidized/reduced ratio of its terminal enzyme in order to transfer electrons from the respiratory pathway.

This competition between the two bacteria results in the following outcomes: one of the species sets a lower potential and out-competes the other; or, they both achieve the same potential, and co-exist. The resistance of the "circuit" used by the bacterium to transfer electrons (*i.e.,* the nanowires or say the potentials of the electron shuttles) also affects the final potential that the bacterium can achieve, and thus this is a factor in competition as well. As bacteria are pushed out to further distances from the anode, as they grow and new cells are formed, the length of the wire or connection to the anode surface will grow more distant, and thus, these bacteria will begin to compete less effectively for electrode space (*i.e.*, setting electrode potential). These bacteria will therefore die, be forced to move (relocate) to a new more advantageous location on the electrode, or switch metabolism to some other kind of electron acceptor (or switch to a fermentation metabolism). It is perhaps not coincidental that iron-reducing bacteria such as *Geobacter metallireducens* are motile when in search of an insoluble electron acceptor such as Fe(III), but that the flagellum is not expressed once the cells have adhered to a particulate iron surface (*Childers et al.* 2002).

3.5 Voltage generation by fermentative bacteria?

Bacteria that produce energy from substrate fermentation obtain energy by substrate level phosphorylation, taking a substrate (such as glucose) and producing a variety of different end products (Fig. 3.3). When a microbe makes acetate, it can produce the most ATP from fermentation. However, it also produces NADH which must be converted back to NAD^+ to sustain the reaction. One possible way for cells to regenerate NAD^+ from NADH is using a reversible hydrogenase that produces hydrogen. Let us compare the potential of the NADH/NAD^+ couple to that of the hydrogen couple ($2H^+/H_2$) under the same conditions (pH = 7), where the hydrogen couple has a potential of $E_0^{/} = -0.421$ V. Thus, under these "standard conditions" (1 M concentration of each soluble species, and 1 bar pressure for H_2), NADH cannot transfer these electrons through a hydrogenase to form H_2 as the process is thermodynamically unfavorable ($E = -0.421 \text{ V} - (-0.320 \text{ V}) = -0.09$ V). However, as indicated above the relative concentrations of NADH/NAD^+ can vary within the cell, allowing the potential to change.

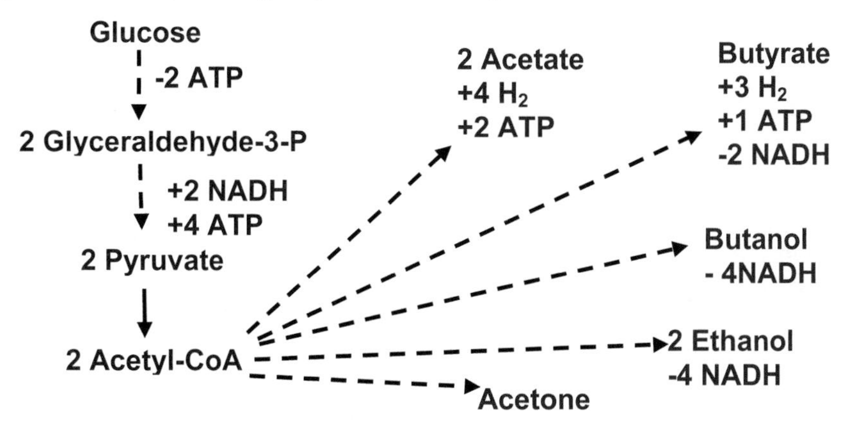

Fig. 3.3 Different fermentation pathways used by *Clostridium acetobutylicum* (ATCC 824) to generate ATP or regenerate NADH (*Girbal et al.* 1995).

In *Clostridium acetylbutylicum*, the build up of NADH shuts down the activity of 2-glyceraldehyde-3-P (*Girbal et al.* 1995). Thus, the cell cannot sufficiently alter the NADH/NAD^+ ratio to make hydrogen production possible as the loss of activity of this enzyme stops the overall reaction. When there is no hydrogen initially present—or if another microorganism (say a methanogen) utilizes the hydrogen—then the reaction becomes favorable. For example, at 10^{-4} bar concentration of H_2, we have

$$E^{0/} = E^0 - \frac{RT}{nF}\ln\frac{H_2}{[H^+]^2} = 0 - \frac{(8.31 \text{J/mol K})(298.15 \text{K})}{(2)(9.65\times10^4 \text{C/mol})}\ln\frac{[10^{-4}]}{[10^{-7}]^2} = -0.296 \text{ V} \tag{3-15}$$

Therefore, the potential for electrons going from NADH to make H_2 at a low concentration of 10^{-2} bar is $E_{emf} = -0.296\ V - (-0.320\ V) = 0.024\ V$. So as long as the concentration of hydrogen remains low, it is possible for electrons to flow from NADH to make hydrogen.

As hydrogen builds up, however, the microorganism must pursue a different metabolic route. Bacteria such as clostridia can make hydrogen using various ferredoxins, enzymes which can have a more negative redox potential than NADH, for example −0.48 V (Table 3.3). Under these conditions, the enzyme/hydrogen couple is now favorable even under standard conditions, or $E_{emf} = -0.421\ V - (-0.480\ V) = 0.059\ V$. Obviously, regulation of the relative concentration of the oxidized and reduced forms of this particular ferredoxin will also allow hydrogen to accumulate to different levels before it becomes thermodynamically unfavorable. It has been found that hydrogen can accumulate to high partial pressures while still allowing hydrogen production, but it has been observed that accumulation of hydrogen does slow down hydrogen production. Continuously releasing the hydrogen pressure, or reducing hydrogen partial pressures through applying a vacuum or gas sparging increases hydrogen yields (*Logan et al.* 2002).

Table 3.3 Properties of several ferredoxins used by different bacteria (T = 303 K, pH = 7) (*Gottschalk* 1986)

Microorganism	Molecular weight (x 10^{-3})	E_0' (V)
Clostridium pasteurianum	6	-0.390
C. acidiurici	6	-0.430
Chromatium vinosum	10	-0.480
Desulfovibrio gigas	18	-0.455
Escherichia coli	12	-0.380
Azotobacter vinelandii	21	-0.225

Implications for electricity generation in MFCs. As was noted in Chapter 2, it has been observed that conductive appendages connected a fermentative and methanogenic bacterium. What would be the thermodynamic advantage of such an association? Obviously, if electrons could be directly transferred from NADH, or a ferredoxin or other electron shuttling compound, there would be no hydrogen gas accumulation in the system. Thus, it would be possible to completely avoid inhibition of metabolism of the fermentative organism by the accumulation of H_2. Moreover, the transfer of electrons from a high potential into a microbe such as a methanogen, which reduces CO_2 to form methane, would have the advantage of providing energy to this microorganism directly. Thus, there could be an advantageous (*i.e.,* syntrophic) association between the two microorganisms.

There is so far little exploration of the ability for fermentative bacteria to participate in *interspecies electron transfer*, in a manner analogous to the well known interspecies hydrogen transfer. The conductive appendages observed between *Pelotomaculum thermopropionicum* and *Methanothermobacter thermoautotropicus* are indeed the first observation of such a situation (*Gorby et al.* 2006), but it is not conclusive that electron transfer is occurring—just that the potential for electron transfer is there. Indeed, an earlier publication on this co-culture provided evidence of why such appendages would be advantageous to the two microbes by keeping them sufficiently close to take advantage of the concentration gradient produced during hydrogen production by *P. thermopropionicum*. A more direct example of the potential for interspecies electron transfer is provided by the observation that isolate EG3, which was closely related to *Clostridium butyricum* was capable of electricity generation in an MFC (*Park et al.* 2001). This finding provides evidence that fermentative bacteria can be iron reducing, and can produce electricity in an MFC. Thus, it is likely that such bacteria could participate in interspecies electron transfer.

These results suggest that interspecies electron transfer may be occurring between fermentative and other microorganisms. Additional research in this area is needed to more fully investigate the potential for this mechanism of syntrophic growth by bacteria.

CHAPTER 4

Power Generation

4.1 Calculating power

To make MFCs useful as a method to generate power, it is essential to optimize the system for power production. Power is calculated from a voltage and current as $P = IE$. The power output by an MFC is calculated from the measured voltage, E_{MFC}, across the load and the current as

$$P = I E_{MFC} \tag{4-1}$$

The current produced by a laboratory-scale MFC is calculated by measuring the potential across the load (*i.e.*, the external resistor, R_{ext}), and using $I = E_{MFC} / R_{ext}$. Thus, we can calculate power output as

$$P = \frac{E_{MFC}^2}{R_{ext}} \tag{4-2}$$

Based on the relationship $I = E_{MFC} / R_{ext}$, we can alternatively express power output in terms of the calculated current as

$$P = I^2 R_{ext} \tag{4-3}$$

Power output normalized by surface area. Knowing how much power is generated by an MFC does not sufficiently describe how efficiently that power is generated by the specific system architecture. For example, the amount of anode surface area available for microbes to grow on can affect the amount of power generated. Thus, it is common to normalize power production by the surface area of the anode, A_{an}, so that the power density produced by the MFC is

$$P_{An} = \frac{E_{MFC}^2}{A_{An} R_{ext}} \tag{4-4}$$

The surface area used for calculating A_{An} is not the same in all studies. In reactors where the anode is suspended in water, for example, the area is defined as a projected or geometric surface area based on both sides of the electrode, *i.e.*, $A_{An} = 2l_{An} w_{An}$ where l_{An} and w_{An} are the length and width, respectively, of a rectangular shaped electrode (*Min et al.* 2005a). When the anode is pressed onto a surface, however, only one side of the anode may be used (*Liu et al.* 2005b; *Liu and Logan* 2004). In other cases, the total surface area of the electrode, *i.e.,* the surface area that includes all accessible surface (*Ringeisen et al.* 2006). When total surface areas are calculated using gas adsorption data (using Argon for example), the useful amount of surface area will be overestimated as surfaces with pore sizes much less than a micron might be inaccessible to the whole microbe (although nanowires or mediators could gain access, depending on the specific surface geometry). Comparison of anode materials on the basis of projected surface area provides a direct method for determining superior anode materials (as long as power is limited by anode surface area, and not by other factors that affect internal resistance, as noted below).

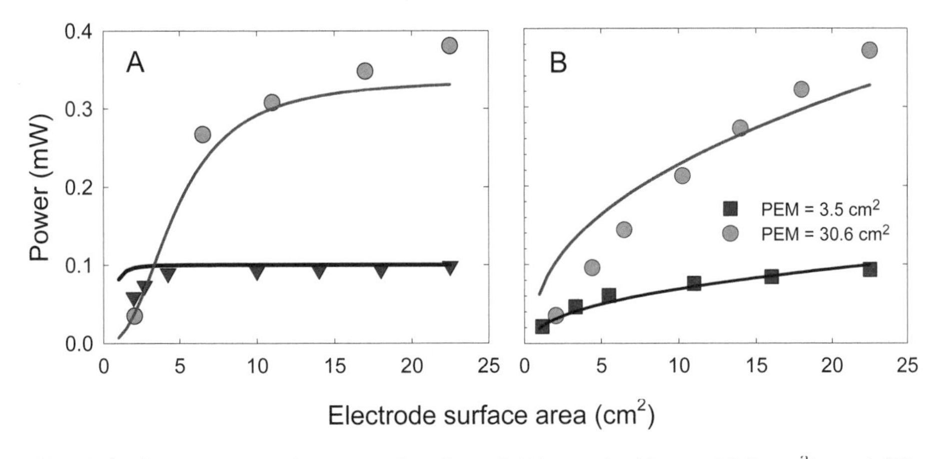

Fig. 4.1 Power generation as a function of (A) anode (A_{Cat} = 22.5 cm^2), and (B) cathode surface area (A_{An} = 22.5 cm^2) for A_m = 3.5 cm^2 (R_{ext} = 1000 Ω) or A_m = 30.6 cm^2 (R_{ext} = 178 Ω). The circuit resistance was chosen so that it did not limit power output (Pt-coated carbon cathode with dissolved oxygen). The lines shown in the figure were predicted using an empirical fit of the data from several experiments. [From *Oh and Logan* (2006). Reprinted with kind permission from Springer Science and Business Media]

As we will see in later examples, the surface area of the anode does not always affect power production. Therefore, in some systems where the anode is moved around in the reactor (*Cheng et al.* 2006b), or systems with very high anode surface areas relative to the cathode area, it is useful to normalize power production by the cathode surface area, A_{Cat}, or

$$P_{Cat} = \frac{E_{MFC}^2}{A_{Cat} R_{ext}} \tag{4-5}$$

In systems that contain a membrane separating the two electrode chambers, it is also possible to normalize power based on the membrane projected surface area, A_m (*Oh and Logan* 2006). This allows an analysis of how relative electrode surface areas affect power while keeping a constant basis for the reactor system. Oh and Logan (2006) determined that power production varied depending on the relative sizes of the anode, cathode and PEM, and suggested that the relative performance of the system be evaluated with equally sized anodes, cathodes and membranes (see Fig. 4.1).

Power output normalized by volume. As MFCs are designed to maximize total system power, ultimately the most important factor is the power production on the basis of the total reactor volume. This is calculated as

$$P_v = \frac{E_{MFC}^2}{v R_{ext}} \tag{4-6}$$

where P_v is the volumetric power (W/m^3) and v the total reactor volume (*i.e.,* the empty bed volume). The reactor liquid volume can also be used, but the convention in environmental engineering is to use the total reactor volume. Sometimes researchers normalize the power production to the anode volume or the anode liquid volume, but these approaches neglect the fact that both the electrode chambers contribute to the overall reactor size. While a feed bottle for the reactor is not included in calculating the reactor size, if cells are grown outside the reactor this volume should be included as well. For example, Ringeisen *et al.* (2006) produced 500 W/m^3 in a small (1.5 cm^3) reactor, but they did not include the 100 cm^3 reactor used to grow and maintain cells pumped into the reactor. An air-cathode system does not have any "space" associated with the air on the outside of the reactor, but if cells are stacked together there must be spacing between the liquid chambers for the air side of the cathode.

Power output compared to total power—the effect of internal resistance and the OCV. Why is it that some MFCs produce only a few milliwatts per reactor volume, while others produce hundreds of watts? The main reason is the internal resistance of the reactor compared to the intrinsic maximum possible potential due to the chemical reactions at the anode and cathode (*i.e.,* the cell E_{emf}). The total amount of power produced is not only due to power lost to the internal resistance of the system. We can view the MFC as having current through two resistors linked in series, with one being the external load (*i.e.,* R_{ext}) and the other the internal resistance, R_{int}. The total maximum power theoretically possible is

$$P_{t,emf} = \frac{E_{emf}^2}{(R_{int} + R_{ext})} \tag{4-7}$$

Thus, we see that the power increases in proportion to the square of the maximum potential. As explained in the previous chapter, however, the *OCV* will always be less

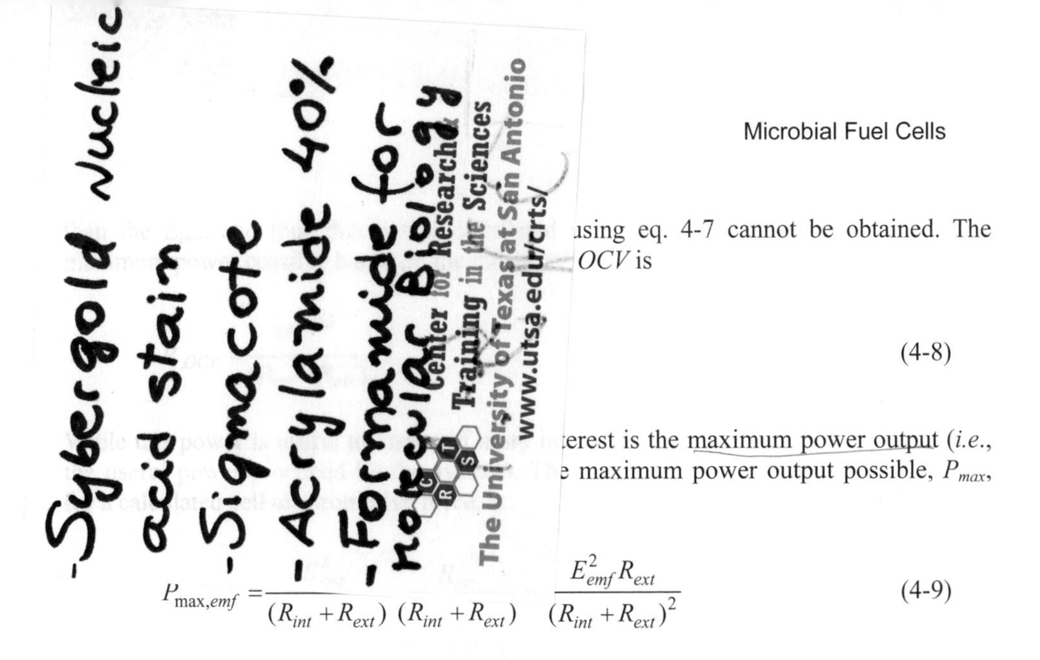

…using eq. 4-7 cannot be obtained. The … OCV is

(4-8)

…terest is the maximum power output (*i.e.*, … e maximum power output possible, P_{max},

$$P_{\max,emf} = \frac{\quad}{(R_{int}+R_{ext})\,(R_{int}+R_{ext})} \quad \frac{E_{emf}^2 R_{ext}}{(R_{int}+R_{ext})^2} \tag{4-9}$$

As noted above, however, the OCV is a more useful measure of maximum power. Thus, we can consider here the most useful relationship for maximum power based on the measured OCV is

$$P_{\max} = \frac{OCV^2\, R_{ext}}{(R_{int}+R_{ext})^2} \tag{4-10}$$

For an air cathode MFC, we find the OCV does not vary appreciably, and thus the main factor affecting power is R_{int}. From eq. 4-10, we see that power if $R_{int} = R_{ext}$, then $P_{\max} = E_{emf}/4R_{int}$. Thus, the maximum power is produced for the smallest internal resistance. Therefore, our objective in MFC construction is to minimize the internal resistance. Methods to measure, understand and improve the internal resistance are discussed below.

Example 4.1

Compare the maximum possible power density (mW/m^2) as a function of current density (mA/cm^2) in an MFC with an air cathode having a surface area of 4.9 cm^2, assuming an OCV of 0.8 V, for $R_{int} = 50\ \Omega$ or $R_{int} = 100\ \Omega$.

We can use eq. 4-10 for calculating the power possible in the system, but we need to use the OCV. For the case of $R_{int} = R_{ext} = 50\ \Omega$, and we have

$$P_{\max} = \frac{OCV^2\, R_{ext}}{(R_{int}+R_{ext})^2} = \frac{(0.8\ \text{V})^2(50\ \Omega)}{(50\ \Omega + 50\ \Omega)^2} = 0.0032\ \text{W}$$

If we now include the surface area of the cathode, and convert to the requested units, we have:

$$P_{\max,Cat} = \frac{P_{\max}}{A_{Cat}} = \frac{(0.0032\ \text{W})}{(4.9\,\text{cm}^2)}\,\frac{10^3\ \text{mW}}{\text{W}}\,\frac{10^4\ \text{cm}^2}{\text{m}^2} = 6530\frac{\text{mW}}{\text{m}^2}$$

To calculate the current, we use $P = I^2R$, but we must be sure to use current (not normalized to area). After rearranging, and with the area correction, we have

$$I_{Cat} = \frac{(P_{\max} R_{int})^{1/2}}{A_{Cat}} = \frac{[(0.0032\ \text{W})(50\ \Omega]^{1/2}}{(4.9\ \text{cm}^2)} \frac{10^3\ \text{mA}}{\text{A}} = 82\ \frac{\text{mA}}{\text{cm}^2}$$

Following the same approach for $R_{int} = 100\ \Omega$, we calculate a maximum power of 2900 mW/m^2. Comparing these internal resistances over a wider range of external resistances (up to 300 Ω), we have the figure shown below. We can see that the maximum power indeed occurs when the internal and external resistances are equal.

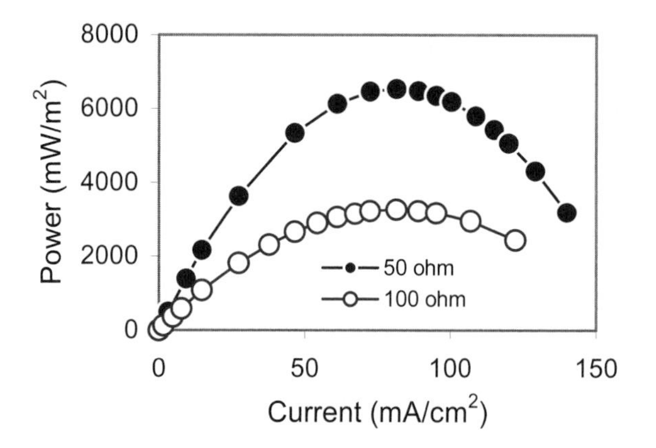

Note that these power densities and current densities are much higher than those presently achieved in MFCs. The reason the power density is so large here is that we are assuming we obtain 0.8 V at the point of maximum power. Typically, we only obtain this high a voltage under open-circuit conditions.

4.2 Coulombic and energy efficiency

While generating power is a main goal of MFC operation, we also seek to extract as much of the electrons stored in the biomass as possible as current, and to recover as much energy as possible from the system. The recovery of electrons is referred to as *Coulombic efficiency*, defined as the fraction (or percent) of electrons recovered as current versus that in the starting organic matter. As we saw in Chapter 2, the oxidation of a substrate occurs with the removal of electrons, with the moles of electrons defined for each substrate (b_e) based on writing out a half reaction. For acetate, complete oxidation requires b_e = 8 mol-e$^-$/mol, while for glucose it is b_e = 24 mol-e$^-$/mol. Coulombic efficiency, C_E, is defined as

$$C_E = \frac{\text{Coulombs recovered}}{\text{Total coulombs in substrate}} \tag{4-11}$$

An ampere is defined as the transfer of 1 Coulomb of charge per second, or 1 A = 1 C/s. Thus, if we integrate the current obtained over time we obtain the total Coulombs transferred in our system. The C_E can therefore be calculated for a fed-batch system as (*Cheng et al.* 2006b; *Logan et al.* 2006)

$$C_E = \frac{M_s \int_0^{t_b} I\,dt}{F\, b_{es}\, v_{An}\, \Delta c} \tag{4-12}$$

where Δc is the substrate concentration change over the batch cycle (which is usually assumed to go from c_0, the starting concentration, to completion for defined substrates such as acetate, or $\Delta c = c_0 - c = c_0 - 0 = c_0$) over a time = t_b, M_s is the molecular weight of the substrate, F = Faraday's constant, and v_{An} is the volume of liquid in the anode compartment. For complex substrates, it is more convenient to use COD as a measure of substrate concentration, and therefore the C_E becomes

$$C_E = \frac{8 \int_0^{t_b} I\,dt}{F\, v_{An}\, \Delta COD} \tag{4-13}$$

where 8 is a constant used for COD, based on $M_{O2} = 32$ for the molecular weight of O_2 and $b_{es} = 4$ for the number of electrons exchanged per mole of oxygen. For continuous flow through the system, we calculate C_E based on substrate concentration or COD, and the flowrate, q, as

$$C_E = \frac{M_s\, I}{F\, b_{es}\, q\, \Delta c} \tag{4-14}$$

$$C_E = \frac{8\, I}{F\, q\, \Delta COD} \tag{4-15}$$

The *energy efficiency* of an MFC is based on energy recovered in the system compared to the energy content of the starting material. The energy efficiency, η_{MFC}, is the ratio of power produced by the cell over a time interval t divided by the heat of combustion of the organic substrate, or

$$\eta_{MFC} = \frac{\int_0^{t} E_{MFC}\, I\, dt}{\Delta H\, n_s} \tag{4-16}$$

where ΔH is the heat of combustion (J mol^{-1}) and n_s is the amount (mol) of substrate added. While values of ΔH are easily obtained from reference texts for specific compounds, values for actual wastewaters are generally not known. Reported η_{MFC} values for MFCs range from 2% to 50% when easily biodegradable substrates are used (*Liu and*

Logan 2004; *Rabaey et al.* 2003). For comparison purposes, the electric energy efficiency for thermal conversion of methane is <40%.

Example 4.2

In an MFC fed-batch experiment it is found that a current of 0.2 mA/cm^2 (anode surface area of 7.1 cm^2) is produced on average over a 35-hour cycle from acetate (1 g/L) in an MFC with a volume of 28 mL. What is the Coulombic efficiency? Assume the acetate is completely degraded.

For this case, we can simplify the integral in eq. 4-13 as just the product I t_b. For acetate, 8 electrons are transferred per mole and the molecular weight is M = 59 g/mol. The C_E is therefore calculated as:

$$C_E = \frac{M_s\, I\, t_b}{F\, b_{es}\, v_{An}\, \Delta c} = \frac{(59\,\text{g/mol})(0.2\,\text{mA/cm}^2)(35\,\text{h})(7.1\,\text{cm}^2)}{(96{,}500\,\text{C/mole}^-)(8\,\text{mol e}^-/\text{mol})(0.028\,\text{L})(1\,\text{g/L})} \frac{1\text{A}}{10^3\,\text{mA}} \frac{3600\,\text{s}}{\text{h}} = 0.49$$

Thus, we see the Coulombic efficiency is about 49% over the batch cycle.

4.3 Polarization and power density curves

The *OCV* measured for an MFC is the maximum voltage that can be obtained with the system, with the limitations imposed by the specific bacterial community and the obtained *OCP* of the cathode. For an MFC, as with any power source, the objective is to maximize power output and therefore to obtain the highest current density under conditions of the maximum potential. The *OCV* is only achieved under a condition where there is infinite resistance. As we reduce that resistance, we lower the voltage. Thus, we look to have the smallest possible drop in voltage as the current is increased in order to maximize the power production over the current range of interest.

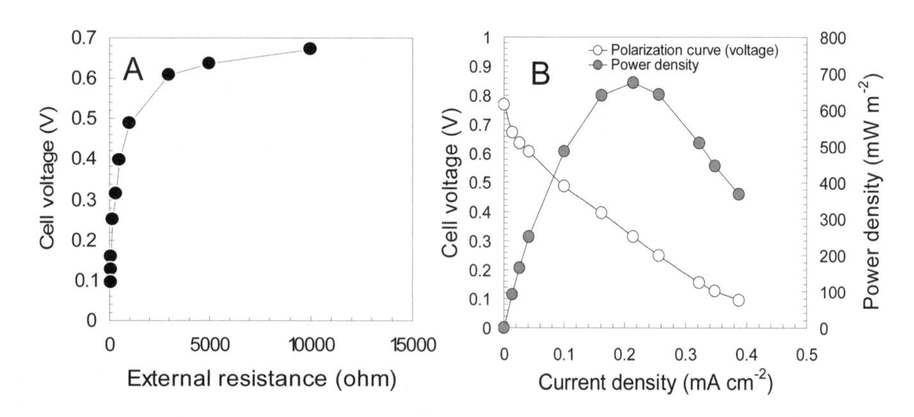

Fig. 4.2 Polarization and power density curves. (A) By switching out the circuit load (external resistance), we obtain a data set on the cell voltage as a function of resistance. Using these data, we calculate (B) the current and plot voltage versus current or current density to obtain the polarization curve; and the power, to obtain the power density curve.

A polarization curve is used to characterize current as a function of voltage. By changing the circuit external resistance (load) we obtain a new voltage, and hence a new current at that resistance. Therefore, to obtain a polarization curve we use a series of different resistances on the circuit, measuring the voltage at each resistance, as shown in Fig. 4.2A. We then calculate the current as $I = E/R_{ext}$, or the current density normalizing by an electrode surface area (usually the anode), and plot voltage versus current to obtain the polarization curve. This curve shows us how well the MFC maintains a voltage as a function of the current production. Note in the example shown in Fig. 4.2B that the *OCV* is 0.78 V (no current, infinite resistance) and that the voltage falls quickly to 0.5 V at a current density of 0.1 mA/cm^2, and then linearly with current density past that point.

The power density curve is calculated from the measured voltage as $P = E^2/R$, or alternatively as $P = I^2R$. Be careful when calculating power not to use a surface normalized current (*i.e., I* divided by electrode surface area) as the power will then be calculated incorrectly when the term is squared.

MFC researchers typically use the top of the power curve to report the "maximum power", which for this case shown in Fig. 4.2B would be 700 mW/m^2. When reporting polarization and power densities it is important to include the *OCV* and show a complete curve up to the maximum power, and then include a few points to the right of the maximum power to fully establish the peak in the power density curve.

4.3.1 Factors that affect the cell voltage

The maximum cell voltage for an air-cathode MFC is the $E^{0/}_{cell}$ = 1.1 V, which is calculated for the acetate–oxygen couple described in Chapter 3 (Table 3.1), as shown in Fig. 4.3. The *OCV* produced by the MFC is always less than that predicted by the maximum potential calculations for the cell. So far, a maximum *OCV* of ~0.8 V has been obtained using oxygen at the cathode, with a maximum of 0.83 V obtained (*Cheng and Logan* 2007). As explained in Chapter 2, the difference between anode $E^{0/}$ and the OCP_{An} is likely due to the maximum potentials that can be established by the bacterial enzymes. For the cathode, this difference is a consequence of the potential that can be established with oxygen reduction.

We can see in Fig. 4.3 that there are three characteristic regions of voltage decrease in the MFC: (1) a rapid voltage drop as current flows through the circuit (at high external resistance); (2) a nearly linear decrease in voltage; (3) a second rapid voltage decrease at high current densities. The cell voltage produced at any specific current is considered to be the result of voltage losses due to electrode overpotentials and ohmic losses, or

$$E_{emf} = E^0 - (\Sigma OP_{An} + |\Sigma OP_{Cat}| + I\,R_{\Omega}) \tag{4-17}$$

where ΣOP_{An} and $|\Sigma OP_{Cat}|$ are the overpotentials of the anode and the cathode respectively, and the IR_{Ω} term includes all ohmic losses which are proportional to the generated current (I) and ohmic resistance of the system (R_{Ω}). Overpotentials of the electrodes (*i.e.,* voltage losses) are most evident at low current densities where the voltage rapidly decreases, but it must be recognized that their magnitude at any specific point is current dependent (*i.e.,* these overpotentials change with the specific current). Electrode overpotentials are thought to arise from three basic losses: (*i*) activation; (*ii*) bacterial metabolism; and (*iii*) mass transport.

Activation losses are due to energy lost (as heat) for initiating the oxidation or reduction reactions, and the energy lost through the transfer of an electron from the cell terminal protein or enzyme to the anode surface (*i.e.*, the nanowire, mediator, or terminal cytochrome at the cell surface). These losses are especially apparent at low current densities (*i.e.,* the first region in Fig. 4.3). They can be reduced using improved catalysts at the cathode, different bacteria on the anode, or by improving electron transfer between bacteria and the anode.

Voltage losses due to *bacterial metabolism* are inevitable as these losses are a consequence of bacteria deriving energy from substrate oxidation. In principle, bacteria need sufficient energy only to pump one proton across a membrane using the CAC cycle (see Chapter 2), or to make 1 ATP via a method of substrate phosphorylation. However, unless bacteria are engineered (or newly discovered) to function in this manner of limited proton production per substrate oxidized, most bacteria that use the CAC cycle are able to pump many protons across the membrane per NADH oxidized. Similarly, bacteria that ferment substrates can make several ATP from acetate oxidation. From our perspective of maximizing recovered power in the MFC, we therefore wish to make the anode potential as negative as possible but still permit electron transfer.

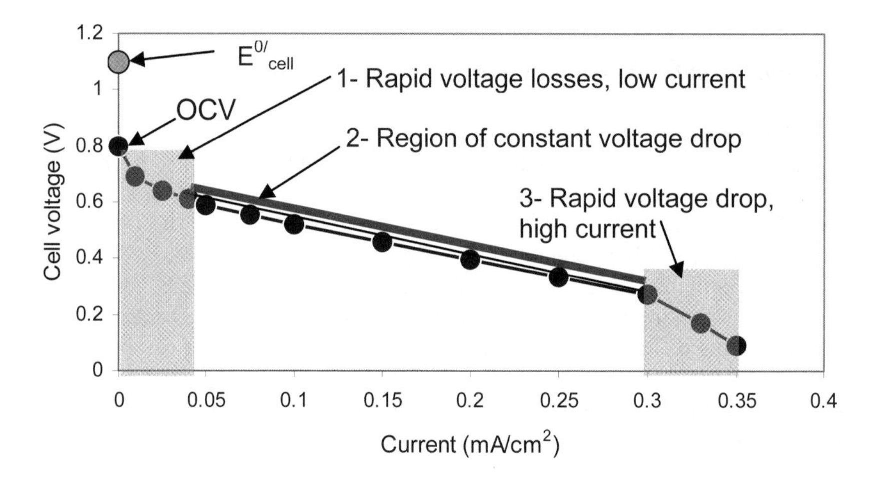

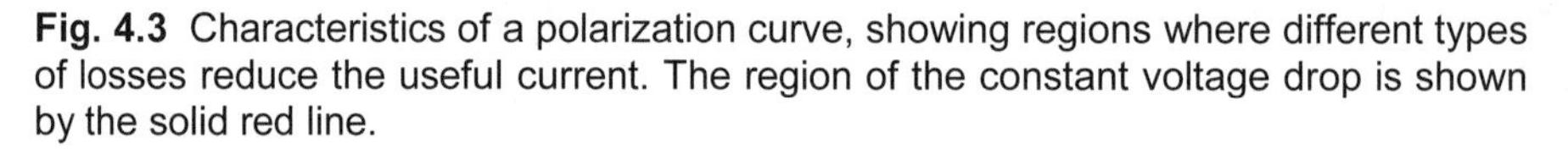
Fig. 4.3 Characteristics of a polarization curve, showing regions where different types of losses reduce the useful current. The region of the constant voltage drop is shown by the solid red line.

Concentration, or *mass transfer losses*, arise when the flux of reactants to the electrode or the flux of products from the electrode are insufficient and therefore limit the rate of reaction. At the anode, the substrate flux to the anode has yet to be an obvious problem in MFC operation (see Chapter 6) as there is little evidence that we have achieved maximum possible power densities based on substrate flux to a surface. However, proton flux from the anode can be a problem as proton accumulation will lower the local pH, adversely affecting bacterial kinetics. Increases in pH have been observed in the bulk fluid near the anode (*Kim et al.* 2007b), and pH within the biofilm could be even lower. Mass transfer limited proton transfer to the cathode can also limit power generation, and result in elevated pH at the cathode (*Kim et al.* 2007b). It is important therefore to maintain sufficient buffer capacity in the system and to minimize the build up of material (and perhaps bacteria?) on the cathode that can hinder proton diffusion to the cathode. There can be bulk phase mass transfer resistances (*i.e.*, diffusion of protons through the solution and membrane if present), but these are evidenced as ohmic resistances as described below.

Ohmic losses. These losses are the most important losses to be overcome for optimum design of the MFC architecture. These losses arise from resistance of ion (proton) conduction due to the solution and (if present) the membrane, and the flow of electrons through the electrode to the contact point (*i.e.*, where the electrodes connect to a wire), and any relevant internal connections. Ohmic losses can be limited by reducing electrode spacing, choosing membranes or electrode coatings with low resistances (if present), ensuring good contacts between the circuit and electrodes, and increasing solution conductivity and buffering capacity.

The ohmic losses due to the solution conductivity between a reference electrode and a bipolar membrane were calculated by ter Heijne *et al.* (2006) as

$$\Delta V_{\Omega} = \frac{\delta_w I}{\sigma} \tag{4-18}$$

where δ_w is the distance in the water (cm), I the current density (A/cm^2) and σ the solution conductivity (S/cm). Each 1-cm distance in a solution of 2.5 mS/cm at a current density of 0.2 mA/cm^2 contributes 0.08 V of ohmic resistance. Thus we see that keeping electrodes as close as possible, and increasing the solution conductivity, should increase power due to a reduction in ohmic resistance.

The challenge to engineering MFCs for greater power production, however, is that reducing ohmic resistance does not always increase power. For example, if the electrodes are too close power production can decrease despite reduction in the ohmic resistance (*Cheng et al.* 2006b). Bacteria can grow only within certain salinity limits, and thus solution conductivity cannot be increased beyond certain ranges (*Oh and Logan* 2006). Proton exchange membranes, such as Nafion, have low resistance but when bonded to the cathode can reduce power production (*Liu and Logan* 2004). Using anion exchange membranes, which so far have been found to have a higher internal resistance than Nafion, can increase power due to conduction of protons using negatively charged species such as phosphate groups (*Kim et al.* 2007b). Of course the cost of the materials that can be used to minimize internal resistance must also be an important factor in the final MFC design.

4.3.2 MFC internal resistance

The most notable feature of the polarization curve is that over the useful range of current (*i.e.*, above a very low current and prior to maximum power production) there is a direct linear relationship between the voltage produced and the current density (region 2 in Fig. 4.3), which can be expressed as

$$E_{emf} = OCV^* - I\,R_{int} \tag{4-19}$$

where the product IR_{int} indicates the sum of all internal resistance losses in the MFC, which are proportional to the internal resistance of the cell (R_{int}) and the current. Note that the *y*-intercept in Fig. 4.3 is not the true *OCV* due to the non-linear and rapid loss of voltage at low current, and thus we indicate the *y*-intercept in eq. 4-19 as OCV^* to clarify that this is the *OCV* implied by extrapolation of the linear portion of the curve to the *y* axis (but not the true *OCV*). This linear relationship between potential and current is a defining characteristic of MFCs due to relatively high internal resistance. Hydrogen fuel cells, including those where hydrogen is produced by fermentation of substrates by bacteria in the anode chamber, often exhibit non-linear relationships between current and voltage over much of the operational region (*Logan et al.* 2006). The linear response in MFCs makes it easy to identify the internal resistance of the cell, as explained below.

A comparison of eqs. 4.17 and 4.19 shows that ohmic losses (R_Ω) are separate from losses due to electrode overpotentials, but the internal resistance of an MFC (R_{int}) includes overpotentials that vary with current. Thus, while these two terms for ohmic and internal resistances are often used interchangeably, it is important to note they are in fact different. Our inability to so far separate out the overpotential and ohmic losses that contribute to the internal resistance of an MFC make it convenient to use R_{int} to classify the MFC. However, as power production increases, and polarization curves become more non-linear, we will need to be careful to separate our definitions of these resistances.

4.4 Measuring internal resistance

There are several different methods to evaluate the internal resistance of an MFC. These include: polarization slope, power density peak, electrochemical impedance spectroscopy (EIS) using a Nyquist plot, and current interrupt methods. The first two of these methods are very easily done, and provide quick methods for estimating internal resistance. The latter two methods are preferred, but they require use of a potentiostat.

Polarization slope method. From eq. 4.19 we can see that in a plot of current versus measured voltage, the slope is R_{int}. Therefore, as long as the polarization curve is linear the internal resistance is easily obtained over the region of interest from the slope of the polarization curve.

Power density peak method. We saw with eq. 4.10 that the maximum power occurred at the point where the internal resistance was equal to the external resistance. Thus, we can identify the internal resistance by noting the external resistance that produced the peak power output.

Example 4.3

Compare the internal resistance for the polarization and power density curves shown in Fig. 4.2 using (a) the peak power, and (b) the polarization curve. Assume the anode surface area is 7.1 cm^2.

(a) From the graph, we read off a peak power of 680 mW/m^2. Converting this to just power production from the system, we have

$$P = P_{An}\,A_{An} = (680\frac{\text{mW}}{\text{m}^2})\,(7.1\,\text{cm}^2)\frac{\text{W}}{10^3\,\text{mW}}\,\frac{\text{m}^2}{10^4\,\text{cm}^2} = 4.83\times10^{-4}\ \text{W}$$

We now estimate the R_{int} as being equal to R_{ext}, which we can calculate from the current of 0.22 mW/cm^2 obtained from the figure at this maximum power density as

$$R_{ext} = \frac{P}{I^2} = (4.83 \times 10^{-4}\ \text{W}) \left[(\frac{\text{cm}^2}{0.22\ \text{mA}})(\frac{1}{7.1\ \text{cm}^2})\frac{10^3\ \text{mA}}{\text{A}} \right]^2 = 200\ \Omega$$

Thus, we estimate the internal resistance as $R_{int} = 200\ \Omega$.

(b) The second method requires calculating the slope. From the graph, we read the points for current and voltage of 0.1 mA/cm^2 and 0.5 V, and 0.3 mA/cm^2 and 0.2 V. Thus, we have for the slope

$$R_{int} = \frac{\Delta E}{\Delta I} = \frac{(0.2 - 0.5)\ \text{V}}{(0.1 - 0.3)\ \text{mA/cm}^2}(\frac{1}{7.1\ \text{cm}^2})\frac{10^3\ \text{mA}}{\text{A}} = 210\ \Omega$$

We see this second approach yields 210 Ω, which is quite similar to that obtained from the first approach, considering that in both cases the values read off a graph were approximate values.

Electrochemical impedance spectroscopy (EIS). This method is preferred compared to the above two methods as the dynamic response of the system is measured, but a potentiostat with EIS software is needed to obtain data. EIS is based on superimposing a sinusoidal signal with small amplitude on the applied potential of a working electrode. By varying the frequency of the sinusoidal signal over a wide range (typically 10^{-4} to 10^6 Hz), detailed information can be obtained on the system by plotting the measured electrode impedance (*Logan et al.* 2006). Impedance measurements for MFC tests can easily be obtained at the *OCV*. A Nyquist plot is then made using the impedance spectra from the real impedance Z_{re} where it intersects the *x*-axis (imaginary impedance $Z_{im} = 0$) (*Cai et al.* 2004; *Cooper and Smith* 2006; *Raz et al.* 2002) using the potentiostat software.

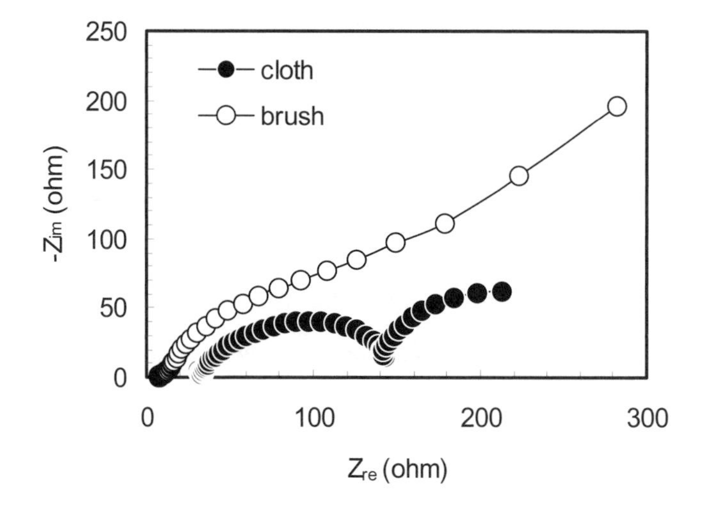

Figure 4.4 Internal resistance obtained from Nyquist plots using impedance spectra of two MFCs with either a carbon cloth or graphite brush electrode. R_{int} is obtained as the value of the *x*-intercept. [Reprinted from *Logan et al.* (2007), with permission from the American Chemical Society.]

An example of a Nyquist plot obtained under these conditions is shown in Fig. 4.4 for a range of 10^5 to 0.005 Hz with sinusoidal perturbation of 10 mV amplitude (*Logan et al.* 2007) The ohmic resistance of the brush electrode was $R_\Omega = 8\ \Omega$, but it was $R_\Omega = 31\ \Omega$ for a carbon cloth electrode (cube MFCs with 4 cm spacing in 200 mM PBS). Additional details of this method can be found in Bard and Faulkner (2001).

Current interrupt method. The current interrupt method requires the use of a potentiostat, and accurate determination of the potential after interrupting the current requires a very fast recording of the potential (μs scale) (*Larminie and Dicks* 2000). The MFC should be operated under steady conditions where there are no concentration losses. To apply the method, the electrical circuit is opened producing zero current (*i.e.*, infinite resistance) and an initial steep rise in voltage (Fig. 4.5). This is followed by a slower and further increase of the potential which will eventually reach the *OCV*. As noted in eq. 4-17, ohmic losses are proportional to the current (*i.e.*, IR_Ω), so when the current is interrupted the ohmic losses instantaneously disappear. This produces the steep rise in the potential (E_R) that is proportional to R_Ω and the current (I) that was produced before the interruption. The ohmic resistance is obtained using Ohms law, or as $R_\Omega = E_R/I$. The electrode overpotentials that occurred during current generation are evidenced by the additional slower increase of the potential (E_A) as it approaches the *OCV*.

Conditions needed for polarization curves or internal resistance measurements. There is not general agreement on conditions that should be established in an MFC prior

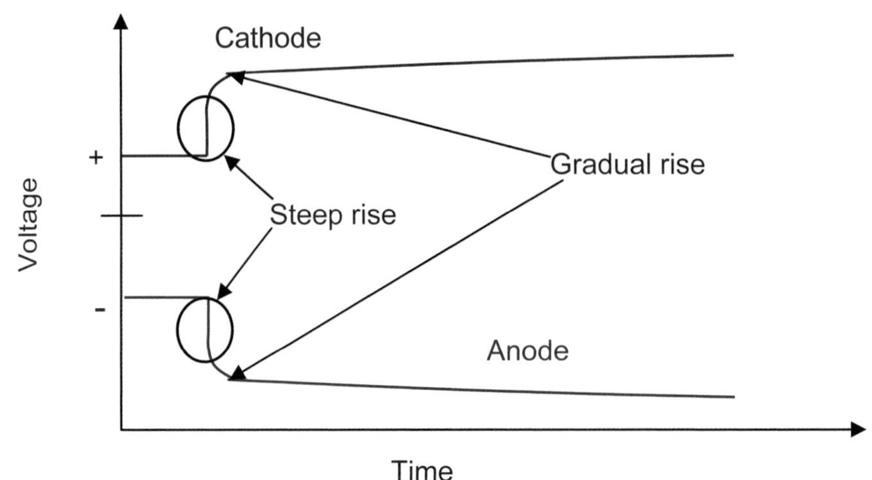

Fig. 4.5 Electrochemical analysis of microbial fuel cells using the current interrupt method. The two responses where there is a steep rise are used for determination of the internal resistance. The subsequent sections with a slower rise indicate voltage losses caused by the activation overpotentials.

to measuring internal resistance or obtaining polarization curves. In some cases, less than one minute (*Finkelstein et al.* 2006) to hours (*Heilmann and Logan* 2006) are used, while in other cases multiple fed-batch cycles are conducted over a period of days (*Cheng and Logan* 2007). The first important criterion before obtaining data is to ensure the system is running under steady state conditions at a set external resistance. For fed batch cycles, this may mean two or three complete cycles that are comparable in power production and duration. For continuous flow systems, there should be good evidence of sustained power over several hydraulic residence times. To vary the circuit load for polarization curves, a potentiostat can be used, but for practical reasons when running a large number of reactors a series of resistors or a variable resistor box is used. When using a potentiostat, an appropriate scan rate is around 1 mV s^{-1} (*Aelterman et al.* 2006).

Prior to obtaining a polarization curve, it is recommended that the MFC be left in the open circuit mode for several hours (or overnight) to fully establish the *OCV*. When the reactor is operated in fed batch mode the best approach is to use a different resistor for one complete cycle (*i.e.,* after feeding until substrate exhaustion). This method is known as the multi-cycle method and multiple fed batch cycles are needed for a complete data set. When tests are being conducted using a reactor with a long batch cycle time (*i.e.,* a week or more), the single-cycle method is used. In this method, the external resistance is varied in a single fed-batch cycle under conditions of maximum power output (*i.e.*, where the substrate has not decreased and affected power output). Data should be obtained for a series of resistances that fully detail the rise to maximum power and the point of maximum power. The resistors can then be switched in order, *i.e.*, from low to high resistance. However, voltages sometimes display hysteresis (*i.e.,* the polarization curves do not overlap in both directions) for reasons which are not well understood. The goal in choosing a time at each resistance is to obtain a pseudo steady-state with each resistor (*i.e.,* the voltage is stable over a reasonable period of time). Times at each resistance reported in the literature vary, with values of 3 minutes (*Menicucci et al.* 2006), 15 minutes (*Logan et al.* 2007), and 7 hours (*Heilmann and Logan* 2006), and therefore there is not a clear indication of the amount of time that should be used. If the time is too short, the system will not equilibrate. If the time is too long, however, it is possible that community structure could change. Only experience with a particular system can help identify the response time characteristics of a particular MFC. In general, while there are differences in the polarization curve depending on times, these can be relatively small compared to differences caused by substrates, extent of acclimation of bacteria, and system architecture. An example comparison of the single- and multiple-cycle methods is in Heilmann and Logan (2006).

4.5 Chemical and electrochemical analysis of reactors

Additional characterization of the reactor and the solution chemistry is needed to better understand and improve the performance of MFCs. The redox potentials within the reactor during substrate removal, as well as localized pH changes around and within biofilms that grow on both electrodes (and the membrane, if present), need to be monitored. One of the major questions that needs to be addressed is, How important are nanowires or mediators? Mediators have conclusively been shown to occur in some studies, but the mechanism of electron transfer is not well understood. Cyclic voltammetry can be used to gain insight into the relative importance of mediators, but the results are often inconclusive. A second important question relates to how the bacteria "fight" for the electrode space, as well as the extent to which they compete to set the potential of the anode (presumably through adjustment of oxidized and reduced states of enzymes as discussed in Chapter 2). These latter two techniques of poised potentials and cyclic voltammetry are discussed briefly below due to their importance in characterizing the MFC bacteria. Presumably, other techniques used in electrochemistry will be applied to the study of MFCs in the near future as well.

Poised potentials. In order to conduct more detailed electrochemical analyses of MFCs, a potentiostat is essential. These can be obtained from various vendors (*e.g.,* Ecochemie, The Netherlands; Princeton Applied Research, USA; Gamry Scientific, USA; Uniscan, UK) and may have one or more channels for monitoring a system. Each channel consists of a three-electrode setup with a reference electrode, a working electrode (anode or cathode), and a counter electrode which is usually Pt. Using a potentiostat, the current or the potential (but not both) in a circuit can be set at a defined value, allowing study of the system under a well controlled condition.

Only one study has been done specifically to examine the effect of a poised potential on the response of the bacterial biofilm. Finkelstein *et al.* (2006) set the potentials of anodes in sediment fuel cells at three potentials meant to mimic different reduction reaction potentials (vs. Ag/AgCl) at the cathode under acetate oxidizing conditions: −0.058 V (arsenate, AsO_4^{3-}/AsO_3^{3-}), 0.103 V (selenite, SeO_3^{2-}/Se^0), and 0.618 V (O_2/H_2O). They found the microorganisms obtained most of the energy (95%) from oxidation of the acetate. When they switched each system to an open-circuit condition, the potentials dropped only 0.040–0.050 V in each case (*i.e.,* they did not all drop to the same value). However, they allowed only 30 s for the system to obtain an *OCV*, and thus the final potential that each reactor would have reached given more time is not known. Still, this study does suggest that the mixed culture adapted in each case to the set potential—as the bacteria would have to do in order to use the electrode as a terminal electron acceptor.

Other studies have been conducted with poised potentials, but usually for the sole reason of excluding oxygen from the system (*Bond et al.* 2002; *Bond and Lovley* 2003; *Chaudhuri and Lovley* 2003). For example, Bond and Lovley (2003) set an anode potential of 0.200 V (vs. Ag/AgCl), resulting in hydrogen gas generation at the cathode. In contrast to the results summarized above, the potential dropped from 0.2 to −0.42 V (vs. Ag/AgCl) when the potentiostat was turned off. This was noted to be equivalent to an anode potential of −0.17 V (NHE) in an open-circuit mode. While using a poised potential is useful at avoiding the need for oxygen at the cathode, setting the potential has two effects here: (1) A net amount of energy is input into the circuit (see Chapter 8 on the MECs for a more detailed analysis of energy input); (2) the creation of hydrogen can result in back-diffusion of hydrogen into anode chamber. The avoidance of oxygen means that obligate anaerobes, such as *G. sulfurreducens* used in this study, will be maintained under anoxic conditions. In mixed culture studies this omission of oxygen could affect community development. The potential for hydrogen cross over into the anode chamber can set up the situation where hydrogen is essentially "recirculated" from the cathode to anode chamber, providing an additional source of energy for the bacteria. This back-diffusion of hydrogen may have contributed to the extremely Coulombic efficiencies (as high as 96.8%) calculated in poised electrode tests (*Bond and Lovley* 2003), when others have typically found maximum values of only ~85%.

Voltammetry studies. Voltammetry has various uses, for example to determine the standard redox potentials of redox active components (*Rabaey et al.* 2005a), examine the electrochemical activity of microbial strains or consortia (*Kim et al.* 2002; *Niessen et al.* 2004; *Rabaey et al.* 2004; *Schröder et al.* 2003), and test the performance of novel cathode materials (*Zhao et al.* 2005). A potentiostat is needed to conduct voltammetry studies, of which there are two basic types. In linear sweep voltammetry (LSV) the potential of the working electrode (anode or cathode) is varied at a certain scan rate (expressed in V s^{-1}) in one set direction. For cyclic voltammetry (CV), the scan is continued in the reverse direction so that the potential is returned to the starting value (Fig. 4.6). CV can be used to determine the presence of electron shuttles, or mediators, produced by bacteria (*Rabaey et al.* 2004). To begin a scan, the reference electrode is placed close to the anode (working electrode), and an appropriate scan rate is chosen. A scan rate of 25 mV s^{-1} has been used for bacterial suspensions (*Park et al.* 2001; *Rabaey et al.* 2004), but for the analysis of mediators in biofilms the scan rate is decreased to 10 mV s^{-1} or less.

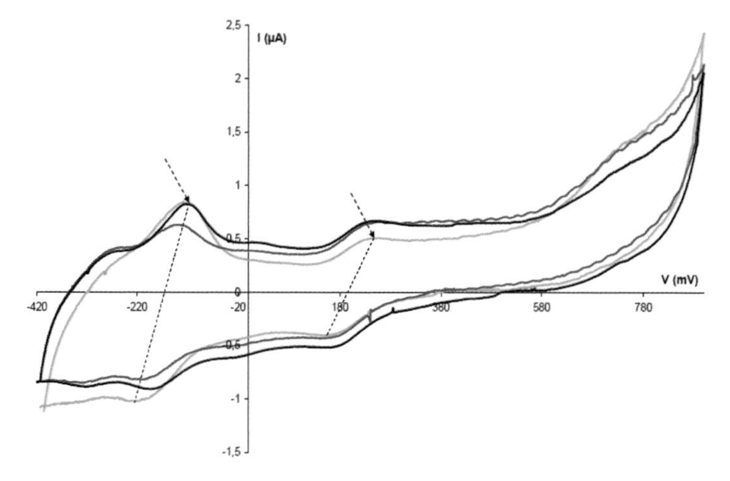

Fig. 4.6 Cyclic voltammograms of a mediator-producing bacterial suspension fed glucose at time 0 (black line), and after 30 min (red line) and 120 min (green line). Arrows indicate peaks while dashed lines that cross between oxidation and reduction peaks intersect the *x*-axis at the formal potential of the redox couple. The peak at −170 mV disappeared when bacteria were washed by centrifugation, but the peak at 180 mV never disappeared. There were no peaks observed from sterile medium. [From *Rabaey et al.* (2004), reprinted with permission of the American Society for Microbiology.]

CV analysis of the MFC should be performed to determine the extent of the mediator production by suspended bacteria and the redox potentials of these mediators. This analysis should be conducted on: (*i*) the MFC-derived cell suspension within its medium before the substrate is consumed; (*ii*) the same, but after the bacteria have completely degraded the substrate; (*iii*) the same culture, washed (by centrifugation), and resuspended in the same physiological solution lacking substrate; and (*iv*) a cell-free suspension, which can be obtained from the supernatant of the centrifuged cell suspension. A peak observed in the first three cases could indicate the presence of a mediator that is cell-membrane associated. This peak should be larger than the case when the substrate is exhausted as the substrate oxidation enhances the peak height. A peak only with substrate in the first case indicates the substrate (or something in the medium) is acting as a mediator. A peak for the first and fourth cases shows that the mediator is freely mobile in solution. The size of the peaks in the CV unfortunately does not provide unequivocal proof of membrane or mobile mediator production as there may be restricted accessibility of the membrane to the working electrode.

It is important to examine CVs of the medium alone (*i.e.,* a sterile solution). Sometimes a medium is prepared with redox indicators, such as reazurin, which are commonly used in anaerobic media. However, redox indicators act as mediators and therefore should not be used in the media. L-cysteine is often used to scavenge dissolved oxygen, but it has also been claimed to act as an electron shuttle under certain conditions (*Doong and Schink* 2002). More importantly, the cysteine is biodegradable and thus can serve as an electron donor for some bacteria (*Logan et al.* 2005). Iron can act as a shuttle, going from the Fe(III) to Fe(II) state by bacterial reduction, but then re-oxidation at the cathode (or membrane) through reaction with oxygen. Ferricyanide will also act as a mediator (*Bond et al.* 2002; *Emde et al.* 1989). A CV test will identify the potential for these exogenous mediators.

Analysis of the biofilm using CV should also be done to determine the potential for biofilm-associated mediators. If a reactor is operated in fed batch mode, and the reactor solution is completely removed at the end of the batch cycle, then soluble mediators will also be removed. This production of mediators that are easily lost would be a poor strategy for optimization of cell growth due to the energy needed to synthesize mediators. However, if the mediators can remain tethered in the biofilm, then they may not be lost through solution changes. There is at present no direct data to support such a biofilm-associated hypothesis. However, there is indirect evidence that suggests mediators may not be very mobile when biofilms are present. Rabaey *et al.* (2005a) extracted phenazine compounds from a pure culture of *Pseudomonas aeruginosa* KRP1. When these compounds were added back into a reactor from which mediators had been removed, power was restored but only to 50% of the original level. Adding a higher concentration of mediators did not produce more power, suggesting that they did not function as originally utilized when extracted and added back to the cell suspension. They further observed that pyocyanin stained cells blue, suggesting that these mediators have a high affinity for the cells which may limit their dissolution into the medium outside the biofilm. The actual functioning of these mediators is quite speculative, however, and additional work is needed to explore the mobility and transport of these mediators in MFC biofilms.

CHAPTER 5

Materials

5.1 Finding low-cost, highly efficient materials

The main challenge in constructing an MFC is first to identify materials and architectures that maximize power generation and Coulombic efficiency, but the next challenge is to minimize cost and create architectures that are inherently scalable. We have already seen, for example, that power densities can be increased by choosing high-energy catholytes such as ferricyanide or permanganate, but the use of these materials is not sustainable or cost effective. To the extent that we can learn about how to better design an MFC, or to manipulate the microbial community for increased power, using such non-sustainable materials is helpful. Ultimately for practical applications we will need to use oxygen as the electron acceptor, and we will need to identify materials which are truly sustainable and scalable. While using bottles and carbon paper electrodes have helped us to understand factors such as internal resistance and the effects (or lack thereof) of different microbes and substrates, our future efforts should be focused on testing materials that are structurally stronger and more practical for achieving large-scale reactors.

The main three components of the MFC are the anode, cathode, and if present, the membrane. The anode has reached the highest level of development with the advent of graphite fiber brush electrodes. Using membranes or materials that may separate anodes and cathodes is a great challenge for MFCs due to their high cost and their general effect of increasing internal resistance. The critical property of a cathode, which is different from the anode, is the need for a catalyst. However, recent research shows several promising approaches for replacing precious metal catalysts with transition metal and other non-precious-metal compounds.

Many of the considerations for choosing materials and deciding on the appropriate MFC architecture are those that apply to any biofilm (fixed bacterial film) bioreactor, and so it is useful to consider briefly here differences and similarities between MFCs and biofilm reactors used for wastewater treatment (WWT). Experience with trickling filter (TF) biofilm reactors, where wastewater flows by gravity over a support material (rocks, random plastic media, or structured plastic media), has resulted in typical engineering specifications for specific surface areas of 100 m^2 of surface area per cubic meter of reactor (100 m^2/m^3) for structured plastic media. Higher specific surface areas are used in more specialized applications, such as nitrifying trickling filters, but in general it has been found that this surface area is ideal for avoiding clogging by biofilms and materials in the wastewater, as well as allowing sufficient air flow through the reactor. Structured plastic media must also be made to be quite strong as it must support the weight of the biofilm and the water. MFCs do not require air flow for the bacteria, but the cathode does require exposure to air. The biofilm in an MFC grows at the structure surface, not at the biofilm–water interface as in a TF. The result is that biofilms in MFCs are likely to have different properties than those in TFs, and so clogging may not be as much an issue in MFCs as in typical TFs. However, bacteria must grow and make new cells, and therefore materials with small pore sizes, like blocks of foam, may not be appropriate as they can fill in and clog.

We can see from this brief comparison that we can expect similarities but also differences in choosing MFC materials and architectures. The different materials, with their respective advantages and disadvantages, are the focus of this chapter. The issue of organizing these materials into architectures, in ways that maximize power and confer an acceptable structure, is addressed in the next chapter.

5.2 Anode materials

The requirements of an anode material are: highly conductive, non-corrosive, high specific surface area (area per volume), high porosity, non-fouling (*i.e.,* the bacteria do not fill it up), inexpensive, and easily made and scaled to larger sizes. Of these properties, the single most important one that is different from other biofilm reactors is that the material must be electrically conductive. A simple test with a voltmeter is sufficient to make the first evaluation of the material by measuring the resistance of the material over a distance. Placing the voltmeter electrodes on a surface, say 1 cm apart, and reading the resistance produces an immediate classification of the conductivity of the material. For example: copper, 0.1 Ω/cm; carbon paper, 0.8 Ω/cm; graphite fiber, 1.6 Ω/cm; carbon cloth, 2.2 Ω/cm; and conductive polymer sheet, 130 Ω/cm. Electrons produced by the bacteria will need to flow from the point of generation on the surface of the material to the collection point (contact with the wire), and only a few ohms of added internal resistance can greatly reduce power. Thus, we need to find materials that are highly electrically conductive—but they must also be non-corrosive, which rules out many metals. In addition, bacteria must be able to attach to the material and achieve good electrical connections. We will therefore see below that some materials such as stainless steel, while apparently meeting many requirements for an anode material, fails to achieve good power generation. Thus, even good conducting materials may not be suitable. We must also consider how the material, or coatings added to the material, affect the ability of the bacteria to transfer electrons (via nanowires, mediators, or by direct contact) to that surface.

Carbon paper, cloth, foams, and RVC. The use of carbon-based electrodes in paper, cloth, and foam forms for the MFC anode is very common. These materials have high conductivity and appear to be well suited for bacterial growth. Carbon paper is stiff and slightly brittle but it is easily connected to a wire (Fig. 5.1A). It should be sealed to the wire using epoxy, with all exposed surfaces of the wire covered or sealed with epoxy as well. Copper wire can be used but it corrodes over time, either releasing copper into solution (which can be toxic to the bacteria) or causing the electrode to detach from the wire. Stainless steel or titanium wires work better in MFCs. Carbon paper is commonly available in plain and wet-proofed versions, with plain paper suggested for anode applications. Carbon cloth is more flexible and appears to have greater porosity than carbon paper (Fig. 5.1B). Carbon foams are much thicker than the cloths, conferring more space for bacterial growth. They have not been as extensively used in MFC studies as the paper and cloth materials. Reticulated vitrified carbon (RVC) has been used in several studies (Fig. 5.1C) (*He et al.* 2005; *He et al.* 2006). The conductivity of the material is excellent at 200 S/cm (5×10^{-3} Ω cm). It is quite porous (97%), with different effective pore sizes specified by a manufacturer. The main disadvantage of the material is that it is quite brittle.

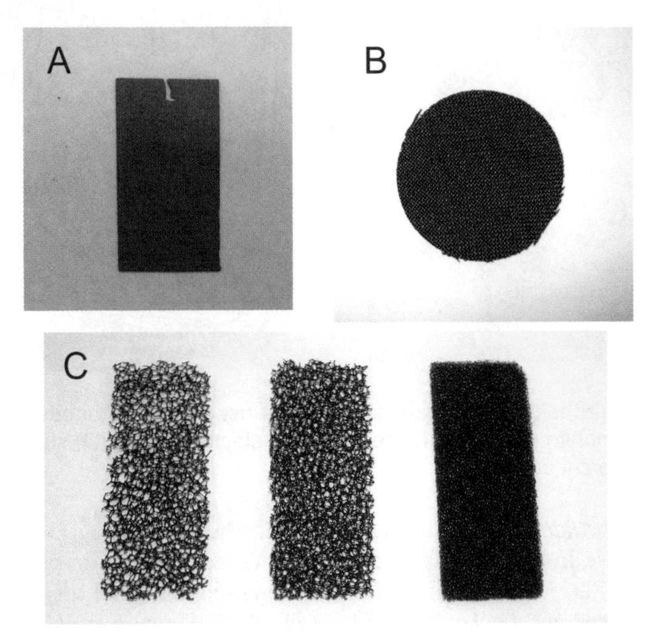

Fig. 5.1 Photographs of carbon materials used for MFC anodes: (A) carbon paper (E-TEK); (B) carbon cloth (E-TEK); (C) three different types of reticulated vitreous carbon (RVC) with different pore sizes (10, 20, and 45 pores per inch). [Photographs: A and B, B.E. Logan; C, L. Angenent].

There are few direct comparisons on the effect of these different carbonaceous materials on power generation. Recall that if high internal resistance limits power generation, increasing anode surface area may not appreciably affect power output, making it difficult to know if one material performed better than another. In two-chambered MFCs with equally-sized electrodes, it was found that an increase in power could be observed when the anode size was increased relative to the cathode only when the CEM size was substantially increased (see Fig. 4.1) (*Oh and Logan* 2006). For example, with CEM and cathode surface areas of A_{CEM} = 3.5 cm^2 and A_{Cat} = 22.5 cm^2, anode sizes of 5 cm^2 or more did not increase power. However, with A_{CEM} = 31 cm^2, power increased with anode size for all anode sizes examined. Thus, it is important to be sure that the anode is limiting power production when investigating new anode materials.

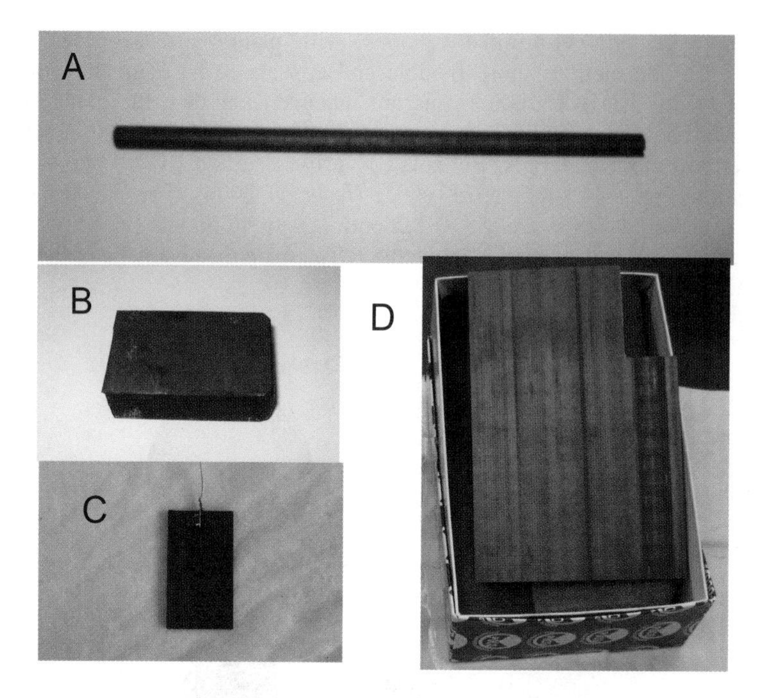

Fig. 5.2 Photographs of some graphite materials used for MFC anodes: (A) graphite rod; (B) thick graphite plate, (C) thinner graphite electrode, and (D) sheet shown with square electrode cut out.

Graphite rods, felt, foams, plates, and sheets. There are a large variety of graphite materials to choose from for MFC electrodes which vary greatly in price, composition, and surface area (Fig. 5.2). Graphite rods have been used in several MFC studies (*Bond et al.* 2002; *Chaudhuri and Lovley* 2003; *Liu et al.* 2004; *Reimers et al.* 2006) as they are highly conductive and have relatively defined surface areas (low internal porosity), and they have been extensively used in electrochemical studies. High-performance graphite rods can be purchases from a number of vendors, but before being used they are often sanded lightly to increase surface area for bacterial growth. A common graphite rod is pencil lead, which is quite conductive (0.2 Ω/cm). Graphite sheets can be purchased in a variety of thicknesses and, like pencil lead, are soft and will mark paper. Because these sheets are flat they are excellent surfaces to use for microscope-based analysis of electrochemically active biofilms. However, graphite sheets are not porous and thus produce less power per geometric (projected) surface area than felts or foams.

Current densities using a graphite rod, graphite felt or graphite foam were compared by Chaudhuri and Lovley (2003). They using found that increasing the total accessible geometrical (projected) surface area increased current generation by *Rhodoferax ferrireducens* in two-chamber, poised-potential MFCs. Graphite felt (A_{An} = 20×10^{-3} m^2) produced three times more current than graphite rods (A_{An} = 6.5×10^{-3} m^2). However, this effect was due to surface area, not difference in materials. When the current was normalized to the surface area the current density and the mass of cells was approximately equal for the two materials (28–32 mA/m^2, and 0.032–0.047 mg-protein/cm^2). The graphite felt produced 2.4× more current, but also held 2.7× more

Fig. 5.3 Photographs of (A) graphite granules 1.5 to 5 mm in diameter; (B) large graphite brush 5 cm diameter and 7 cm long with a specific surface area of 7170 m^2/m^3 of brush volume; (C) smaller graphite brush 2.5 cm in diameter and 2.5 cm in length with 18,200 m^2/m^3 of brush volume; and (D) section of a tow of graphite fibers.

bacteria, likely as a consequence of the higher porosity of the felt compared to the graphite rod. Long-term performance of these materials was not reported.

Graphite granules. Graphite granules (Fig. 5.3A) are chunks of graphite that resemble pencil lead in their appearance and are available from many different sources (*e.g.,* Graphite Sales, Inc., Chagrin Falls, OH). The first use of these materials was reported by Rabaey *et al.* (2005b) as an anode material, but they have since been used in other packed-bed reactors as both anodes and cathodes (*Aelterman et al.* 2006; *Heilmann and Logan* 2006; *Rabaey et al.* 2006). Granule sizes can vary, but so far granules that have been used have reported diameters of d = 1.5 to 5 mm (Le Carbone, Belgium) with reported specific surface areas estimated to range from 820 to 2700 m^2/m^3 and a porosity of 0.53 (*Rabaey et al.* 2005b). This porosity is typical of a loosely packed bed and can be related to specific surface area using $A_s = 6\theta / d$ (*Logan* 1999), so for $\theta = 0.53$ and $d = 3$ mm we have $A_s = 1100$ m^2/m^3. The granules are conductive (0.5–1.0 Ω/granule), but in order to make the complete bed conductive the granules must make good electrical contact with each other. Due to the shape of the granules and the bed porosity, they can connect at only a small fraction of their total surface area.

Graphite fibers and brushes. The highest specific surface areas and porosities for anodes can be achieved using graphite fiber brush electrodes (Fig. 5.3B) (*Logan et al.* 2007). These brushes can be made from carbon fibers produced by different manufacturers (*e.g.,* ZOLTEK) using conventional industrial brush machines. The core of the wire can be made from non-corrosive metal, such as the two titanium wires used for the brush shown above. Due to the small diameter of the graphite fibers (7.2 μm), extremely high specific surface areas can be achieved. The large brush shown above (5 cm diameter, 7 cm long) has an area of 1.06 m^2, producing an estimated specific surface area of 7170 m^2/m^3-brush volume with a 98% porosity (Fig. 5.3B). The smaller brush has

0.22 m^2, or 18,200 m^2/m^3 of brush volume and a porosity of 95% (Fig. 5.3C). The surface area per volume as used depends on the volume of the reactor.

Random fibers can also be used for the anode, but getting good dispersal of the fibers can be a problem. Fibers are sold in units called "tows," or long strands of the fibers. The tow of fibers shown in Fig. 5.3D has a mass of ~0.1 g, with a projected surface area of 0.020 m^2 based on the fiber diameter of 10 μm (Granoc-Nippon). For comparison purposes, the same length of graphite fibers produced by another manufacturer (Fibre Glast, Ohio) had 0.035 m^2 per tow and a single-fiber diameter of 6 μm.

Large brush anodes and random fibers were compared for power generation in single-chamber bottle MFCs (see Chapter 6), although not on a same mass loading basis. The brushes were tested under conditions which produced a specific surface area of 4200 m^2/m^3 of reactor volume, and produced up to 1430 mW/m^2 (2.3 W/m^3, CE = 23%) versus 600 mW/m^2 with a plain carbon paper (23 cm^2, both sides) anode (power normalized to the cathode projected surface area) (*Logan et al.* 2007). The graphite fibers in a random (tow) formation produced up to 1100 mW/m^2. Tests with different masses of fibers did not produce a clear trend of power per mass loading, suggesting that fiber clumping was a factor in the performance of this system.

Testing of the smaller brush electrode in a cube-type MFC produced the highest power density yet achieved for an air cathode MFC of 2400 mW/m^2 at a current density of 0.82 mA/cm^2 (R_{ext} = 50 Ω) (*Logan et al.* 2007). The power was 73 W/m^3 based on normalizing power to the reactor liquid volume. C_E ranged from 40% to 60% depending on current density.

Conductive polymers. The use of conductive polymer materials as anodes has not been well investigated in MFC studies. Work in the author's laboratory by Heilmann (unpublished) showed that polymers were not as effective as carbon paper or granules. Two conductive coatings applied onto polyvinylchloride (PVC) (Baytron F HC and Baytron P HC V4; H. C. Stark; Newton, MA), along with one of the conductive coatings (Baytron P HC V4) on a conductive polymer (TP5813; Premix, Milton, WI), were compared for power production in cube MFCs with 2-cm electrode spacing. The best-performing polymer anode was the P HC V4 coating on PVC, which produced a maximum voltage of 99.4 ± 1.9 mV. However, this voltage was only 55% of that achieved using a carbon cloth anode (181 ± 15 mV). The F HC coating produced *ca.* 17 mV, while the coating on the conductive polymer produced 33–53 mV. Performance of all these polymer systems was erratic and inconsistent, indicating both maximum power densities and stability of these materials will be a concern. However, these were preliminary results and additional work is needed in this area.

Metals and metal coatings. The use of various metals and metal coatings on carbon materials has not been well-examined for MFCs. In one study, it was shown that addition of vapor-deposited iron oxide to a carbon paper cathode decreased acclimation time of a reactor but did not affect maximum power (*Kim et al.* 2005). A two-chamber reactor was used, and thus power production was limited by high internal resistance so it is not known if the iron oxide would have resulted in an increase in power production in a reactor with less internal resistance. Over time the iron oxide coating dissolved into solution, leaving only the carbon paper electrode. However, the use of iron to enrich iron-reducing bacteria on the electrode may be a beneficial approach for acclimation of an electrode with exoelectrogenic bacteria.

Metals added to electrodes have been shown to increase power in several studies, but whether this additional power resulted from galvanic increases due to the potentials of the metals or for other reasons has not been well-examined. Using a self-made Mn^{4+}-graphite anode (graphite, manganese ion, nickel and a binder), Park and Zeikus (2002) produced 10.2 mW/m^2 compared to only 0.02 mW/m^2 with a woven graphite electrode with a pure culture of *Shewanella putrefacians* and lactate as substrate. Additional experiments using the same electrode produced 788 mW/m^2 with complex medium of lactate, peptone and yeast extract and a sewage sludge inoculum compared to 0.65 mW/m^2 using a woven graphite electrode (*Park and Zeikus* 2003). In these latter experiments the CEM was replaced by a porcelain septum made from kaolin. The lack of a traditional CEM may have produced a lower internal resistance, resulting in greater power densities, but no values for internal resistance were provided in their studies and comparisons were not made with the same culture of *S. putrefacians*.

Fe_3O_4 and $Fe_3O_4 + Ni^{2+}$ were added to graphite paste, and $Mn^{2+} + Ni^{2+}$ were added to a graphite-ceramic mixture to make anodes used in sediment fuel cells using the same procedure employed by Park and Zeikus (2002). These anodes produced 1.7- to 2.2-fold greater current than plain graphite electrodes under similar conditions (*Lowy et al.* 2006). The Mn^{2+}-based electrode, which performed best in these laboratory tests, was also field-tested in sediment MFCs and produced 105 mW/m^2, versus ~20 mW/m^2 obtained in previous tests.

Additional work on the addition of metals to electrodes has been done, but this work was not published due to disappointing results. Highly conductive stainless steel frits were tested as anodes, but they produced much less power (0.14 mW/m^2, 1 kΩ resistor) than carbon paper (350 mW/m^2) in the same cube-reactor system (equal-sized electrodes). Additional tests done by adding metals and metal oxide powders to conductive carbon coatings on conductive polymer electrodes have also so far not produced any increases in power. Materials tested included aluminum oxide (Al_2O_3), iron oxide (Fe_2O_3), 304 stainless steel, titanium, and tungsten. Alumina, stainless steel, and titanium inhibited electricity production, while tungsten and iron oxide decreased the start-up times but overall produced less power. Iron oxide was found to produce greater power than MFCs containing only carbon in the anode coating, but these materials on conductive polymers did not exceed the performance of plain carbon paper or cloth electrodes.

Non-metallic treatments and modifications of anodes to increase power. Several approaches have been taken to increase performance of the anode by using different non-metallic materials. The most successful method so far in terms of maximizing power production was achieved using an ammonia-gas-modified carbon cloth electrode. Treatment of carbon cloth using 5% NH_3 gas in a helium carrier gas at 700ºC (for 60 minutes) increased the positive surface charge of the cloth from 0.38 to 3.99 meq/m^2. This reduced acclimation time needed for a wastewater culture to produce power by 50%, and it increased power to 1970 mW/m^2, compared to (a) 1640 mW/m^2 without the ammonia treatment in the same high ionic strength phosphate buffer and (b) 1330 mW/m^2 in a lower ionic strength solution.

Other treatments of the anode have increased power generation relative to controls, but the lower overall performance of these systems (likely due to higher internal resistance) prohibits conclusions about which methods work better than others. Park and Zeikus (2002) bound a known mediator, neutral red (NR), to a woven graphite electrode and increased power to 9.1 mW/m^2 compared to only 0.02 mW/m^2 with the same electrode lacking NR, using a pure culture of *Shewanella putrefacians* and lactate as substrate. In an improved system lacking a CEM, they achieved 845 mW/m^2 with complex medium (lactate, peptone and yeast extract) and a sewage sludge inoculum compared to 0.65 mW/m^2 using the plain electrode (*Park and Zeikus* 2003).

The addition of bound mediators has also been tested in sediment fuel cells. Lowy *et al.* (2006) found that AQDS- and 1,4-naphthoquinone (NQ)-modified anodes produced 1.7 and 1.5 times as much current as plain graphite electrodes. The AQDS-modified electrode produced up to 98 mW/m^2 in field tests, which was much better than the plain graphite electrodes (20 mW/m^2) and only slightly less than the 105 mW/m^2 produced using Mn^{2+}+Ni^{2+}-modified electrodes.

Polyaniline was tested as a method to increase power (*Schröder et al.* 2003), but studies were only conducted on anodes containing Pt where power was generated via hydrogen produced by bacteria fermenting a sugar, and not by exoelectrogens. These types of hydrogen MFC (HMFC) reactors are further discussed in the next chapter.

The addition of compounds to anode surfaces is a developing research area and we are only just beginning to understand the reasons for power increases or decreases. For example, adding metal-containing graphite pastes can increase available surface area and thus increase power for reasons independent of the metal in the coating. Adding positively charged metals or using a high-temperature ammonia gas treatment can make the surface more positively charged and thus facilitate more rapid adhesion of bacteria to the surface. The way in which electrons are transferred from the cell enzymes to the electrode surface is not well understood, and so advances in our molecular-scale understanding of charge transfer may need to wait for further analyses of this critical step. Charge transfer by bacterial cytochromes—for example, between cytochrome *c* and gold electrodes in the presence of various self-assembled monolayers (SAMs) (*Tarlov and Bowden* 1991)—has been an ongoing subject of study for many years. Crittendon *et al.* (2006) recently showed that little current was generated with whole cells of *Shewanella putrefacians* on gold electrodes, but that current generation was substantially improved by adding 11-mercaptoundecanoic acid to form a SAM on the electrode. They reasoned that the SAM enhanced electron transfer because it stabilized the electron carriers from the cell to the electrode surface, acting in a manner analogous to humates present in natural water. Humates are naturally occurring recalcitrant compounds in water that strongly adsorb to surfaces, particularly metal oxide surfaces. This finding is an exciting indication that addition of materials to surfaces may facilitate, on the molecular scale, electron transfer to surfaces.

5.3 Membranes and separators (and chemical transport through them)

In a hydrogen fuel cell (HFC), a membrane is an essential component of the system as it separates the two gases (H_2 and O_2) and provides a method for conducting protons between the two gases. The membrane is therefore referred to as a proton exchange membrane (PEM) (although this acronym is also used for polymer electrolyte membrane). In an MFC, however, the water conducts the protons and thus a membrane is not necessarily needed as a system component. Some HFCs use membrane electrode assemblies (MEAs), where the cathode catalyst is directly bonded to the membrane. Liu and Logan (2004) showed that an MFC lacking a membrane produced more power than an MFC with the membrane (Nafion) bonded to the cathode, indicating that the membrane can adversely effect power generation.

So why are membranes used in MFCs? They are primarily used in two-chambered MFCs as a method for keeping the anode and cathode liquids separate. Ferricyanide or water containing dissolved oxygen in the cathode chamber cannot be allowed to mix with the solution in the anode chamber. These membranes need to be permeable so that protons produced at the anode can migrate to the cathode. A piece of solid foil between the chambers, for example, will not transfer protons. Membranes also serve as a barrier to the transfer of other species in the chamber. For example, they can be used to reduce the unwanted substrate flux from the anode to cathode (fuel crossover), and oxygen from the cathode to anode, improving Coulombic efficiency (*Liu and Logan* 2004). Even in single-chamber MFCs, membranes can be useful to isolate the catalyst from the cathode.

The disadvantages of membranes in MFCs are their high cost and that they decrease system performance. Nafion can cost $1400/m^2$, while a simple cation exchange membrane (CEM) costs far less—for example $80/m^2$ (CMI-7000, Membranes International, Inc.). These high costs would make the use of Nafion prohibitive in large-scale application of MFCs (for example, for wastewater treatment). The adverse effect of the membrane on performance is usually a result of increased internal resistance. If the solution conductance or effective diffusivity of a proton, or chemical species carrying a proton, is reduced by the presence of the membrane then the internal resistance of the system will increase and the power production will be reduced.

Table 5.1 Internal resistance and maximum power density for cation (CEM), anion (AEM), and ultrafiltration (UF) membranes tested in bottle (B-MFCs) and cube (C-MFC) MFCs. [Data from *Kim et al.* (2007b)]

Membrane	Internal resistance (Ω)		Maximum power (mW/m^2)	
	B-MFC	C-MFC	B-MFC	C-MFC
No membrane	1230 ± 44	84 ± 3	---[a]	---[a]
CEM (Nafion)	1272 ± 24	84 ± 4	38 ± 1	514
CEM (CMI-7000)	1308 ± 18	84 ± 2	33 ± 2	480
AEM (AMI-7001)	1239 ± 27	88 ± 4	35 ± 3	610
UF-0.5K	6009 ± 58	1814 ± 15	5 ± 1	---[b]
UF-1K	1239 ± 52	98 ± 5	36 ± 0	462
UF-3K	1233 ± 46	91 ± 6	36 ± 0	---[b]

[a]Not applicable. [b]Not measured.

The internal resistance (R_{int}) attributable to the membrane can be measured by placing different membranes into the reactor and measuring internal resistance as described in Chapter 4. There is generally good agreement between R_{int} measurements and maximum power densities measured in MFCs (Table 5.1). In bottle reactors, where the electrodes are spaced at large distances, R_{int} values are high and there may not be much of an effect of the membrane on overall internal resistance. However, in two-chamber cube MFCs (see Fig. 6.1) where the electrodes are closer spaced and internal resistances are lower, we can more clearly see this relationship between R_{int} and power.

Another effect of the membrane is how it is incorporated into the system, as this can affect overall charge transfer. Liu and Logan (2004) found that hot-welding Nafion to the cathode decreased power compared to a system lacking a membrane (262 vs. 494 mW/m^2). An examination of Table 5.1 shows that R_{int} is not intrinsically affected by the presence of the Nafion itself, as Nafion has high proton conductivity. Thus, it was either some artifact of the hot welding process on the system (perhaps the Nafion was damaged

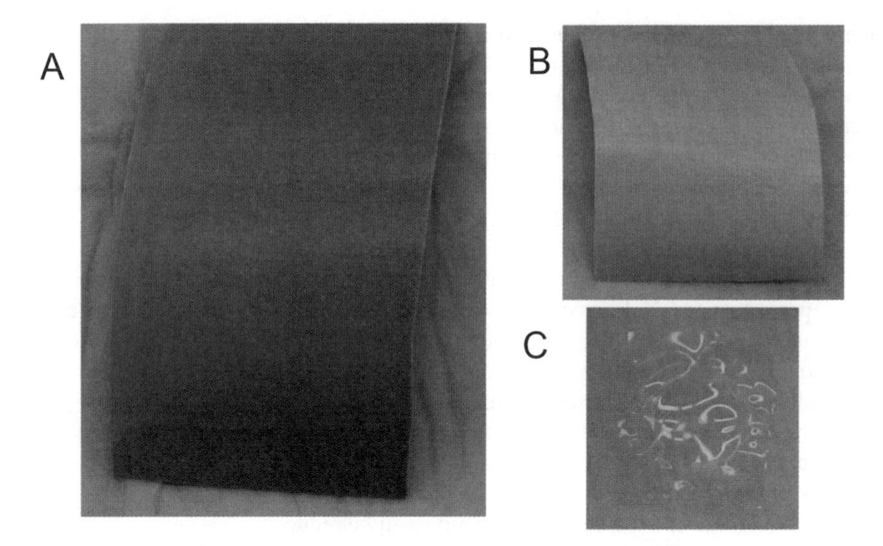

Fig. 5.4 Different membranes tested in MFCs: (A) cation exchange membrane (CMI-7000, Membranes International, Inc.); (B) anion exchange membrane (AMI-7001, Membranes International, Inc.); (C) Nafion 117 membrane (Ion Power, Inc.).

by this process), or this joining process somehow affected proton access (diffusion) to the catalyst. There can be very high, but localized, pH changes that occur near the electrodes which may also have contributed to these differences between the systems with and without membranes (*Rozendal et al.* 2006b).

We now know that cation or anion exchange membranes, or any permeable material, can function as a solution barrier in an MFC if charge can be transferred. The qualities of the membranes used in MFCs, along with their properties, are therefore reviewed in more detail below.

Cation exchange membranes. The most commonly used cation exchange membrane (CEM) is Nafion 117 (Dupont Corp., available from Ion Power, Inc.) (Fig. 5.4). The code 117 is used to distinguish the thickness of the membrane (0.019 cm) from other Nafion membrane thicknesses. This membrane was developed for use in an HFC and thus was optimized to create a stable and conductive environment for high proton concentrations (low pH) under conditions where the water content is carefully controlled. However, this material becomes completely saturated (flooded) with water in an MFC, producing a pH reflective of the solution properties (likely neutral pH). Thus, it does not function according to its intended purpose in an MFC as it cannot operate under its designed conditions.

Nafion is also referred to as a PEM on the basis that it is designed to transfer protons (H^+), but in an MFC it preferentially conducts other positively charged species (Na^+, K^+, NH_4^+, Ca^{2+}, and Mg^{2+}) that are present at typically 10^5 times higher concentrations than protons in solution (*i.e.,* 10^{-7} M at neutral pH) (*Rozendal et al.* 2006b). The competitive transport of cations other than protons significantly affects MFC performance. When substrate is degraded, protons are produced at the anode and consumed at the cathode. If protons cannot migrate at a sufficient rate from the anode to the cathode, the pH will decrease at the anode and increase at the cathode while charge balance is maintained by the migration of other cations. The pH decrease at the anode affects bacterial respiration and thus current generation. At the cathode, the pH can rapidly rise, which can lead to mass transfer limited proton transport to the catalyst. A well-buffered solution can offset these pH changes in a two-chambered MFC, but it is not clear to what extent localized pH changes in the biofilm may affect power generation. Calculations based on oxygen reduction at the cathode show that an increase in pH will affect the cathode potential, as shown in the example below.

A CEM membrane (CMI-7000) made by Membrane International Inc., NJ, has been used in several MFC studies, mostly those with ferricyanide as the catholytes (*He et al.* 2005; *Rabaey et al.* 2005b; *Rabaey et al.* 2003; *Rabaey et al.* 2005c). This membrane is much thicker and stiffer than the Nafion 117 (0.046 cm) and in general appears to be structurally stronger. There are many manufacturers of CEMs which could be used in MFCs, but these have not been compared for their performance in MFC applications.

Anion exchange membranes. If H^+ ions are not effectively transported across a CEM, then how can pH be more effectively balanced in an MFC? Kim *et al.* (2007b) recognized that protons can also be effectively transported by chemicals used as a pH buffer, such as phosphate anions. Therefore, they used an anion exchange membrane (AEM; AMI-7001, Membranes International, Inc.) as the separator in two-chambered MFC tests. They found that higher power was produced using the AEM than with two different CEMs (Nafion and the CMI-7000). By monitoring the phosphate concentrations in the chambers on either side of the membrane, they also showed that phosphate anions were being transported across the membrane and that pH was better maintained in the anode chamber. However, the pH did increase more in the cathode chamber than with Nafion, and thus improvements are still needed in the membrane qualities to improve MFC performance. The higher the current density in a system, the greater the flux of protons (or proton carriers such as phosphate groups) needed to maintain both charge and pH balance.

Bipolar membrane. A bipolar membrane consists of an anion and a cation membrane joined in series. As voltage develops rather than protons passing the membrane water is split, resulting in the transport of anions (OH^-) to the anode and cations (H^+) to the cathode to balance charge. The energy needed for the water splitting reaction is claimed to be small because water is split into ionic species, H^+ and OH^- (not electrolyzed to H_2 and O_2). Ter Heijne *et al.* (2006) developed an MFC based on using a ferric iron catholyte. In order to keep the Fe^{3+} in solution in the cathode chamber, they needed to use a low pH (<2.5), but under these conditions the CEM fouled (see below). By using a bipolar membrane, they were able to maintain the low pH in the cathode chamber and near-neutral pH in the anode chamber. For a completely selective bipolar membrane, it was calculated that 0.83 V was needed to be overcome (*Hurwitz and Dibiani* 2001) for a 100% selective membrane with one molar acid and base solutions (ΔpH = 14) using

$$\Delta V = \frac{-2.3RT}{F}\Delta \text{pH} \qquad (5\text{-}1)$$

The pH difference used in experiments by ter Heijne *et al.* (2006) was pH = 3.5, and thus a voltage drop of −0.21 V was calculated as a result of the pH differences between the two chambers. They concluded that this was sufficient for water splitting, and observations of a drop in catholyte pH seemed consistent with this result as well.

Example 5.1

Compare the effect of pH on the cathode potential for oxygen reduction to water or to hydrogen peroxide (5 mM) over a pH range of 4 to 11.

For oxygen conversion to water, the estimated cathode potential is predicted by eq. 3.6, assuming a partial pressure of oxygen in air of 0.2 atm. For H_2O_2, the equation based on reaction C-3 in Table 3.1 at a pH = 7 is

$$E^{0/} = E^0 - \frac{RT}{nF} \ln \frac{[H_2O_2]}{[O_2][H^+]^2} = (0.695\,V) - \frac{(8.31\,J/mol\,K)\,(298.15\,K)}{(2)\,(9.65 \times 10^4\,C/mol)} \ln \frac{[0.005\,M]}{[0.2][10^{-7}\,M]^2}$$

$$E^{0/} = 0.370$$

Values at other pHs are similarly calculated, with the results shown in the figure below. As we can see, there is a reduction of 0.177 V for either case when the pH changes from 7 to 10. Thus, it is important to maintain neutral or lower pH near the cathode in order to maximize power generation.

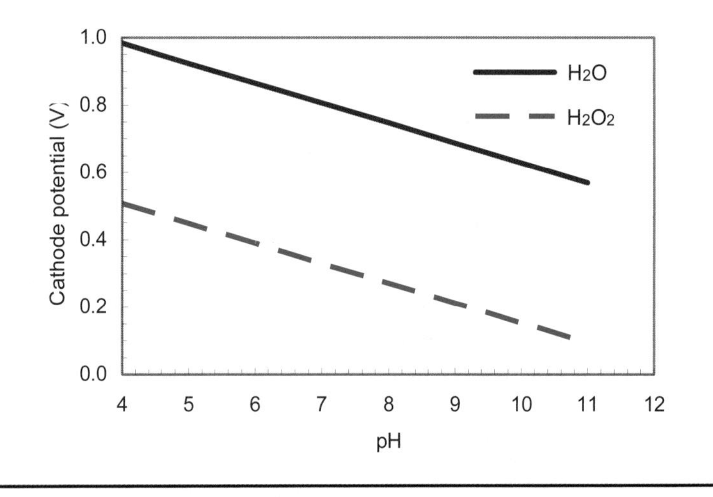

Other separators and membranes. The observation that cations or anions can help to maintain charge in an MFC opens the door to new possibilities for using different types of membranes in MFCs. The main function of the membrane is to keep solutions isolated while allowing charge transfer, presumably via small ions. Thus, ultrafiltration (UF) membranes—especially those developed for wastewater applications—may be suitable for use in MFCs. These membranes have tiny pores, with many membranes available that have molecular weight cutoff values below 1000 daltons (1K). UF membranes have appreciable flow through them only under high pressure, and thus under relatively low hydrostatic pressures in tanks (or when pressures on either side of the membrane are balanced) there is little water bulk flow through the membrane.

Kim *et al.* (2007b) tested three different ultrafiltration membranes (Amicon Corp.) for power generation in two different types of two-chambered MFCs. They found that these membranes had high internal resistances, and thus produced less power than the CEM or AEM membranes (Table 5.1). The 0.5 K membrane had extremely high internal resistance values. As a result, power was only 5 mW/m^2 in two-chamber bottle reactors whereas the other membranes did not appreciably impact power generation as they all produced 33–38 mW/m^2. In the cube reactors, where internal resistance was lower due to a closer electrode spacing and the use of an air cathode, the 1 K membrane produced only slightly less power than the CEM membrane. Thus, in theory it may be possible to replace the CEM membrane with a more conventional ultrafiltration membrane in an MFC, but membranes must be developed that result in lower internal resistances.

Chemical fluxes across membranes (oxygen and substrate). CEMs like Nafion and all other membranes used in MFCs are not only permeable to protons, but also to gases and other chemical species. This characteristic has been most frequently recognized for oxygen diffusion, but it must be recognized that many substrates are able to pass the membrane as well. The chemical flux through the membrane can be calculated in terms of a mass transfer coefficient, $\boldsymbol{k}_{Cm}$, as

$$J_m = \boldsymbol{k}_{Cm}(c_{An} - c_{Cat}) \tag{5-2}$$

where c is the substrate concentration on the anode (c_{An}) or cathode (c_{Cat}) side of the membrane. The chemical diffusivity inside the membrane, D_{Cm}, can be calculated from the membrane thickness, δ_m, as

$$D_{Cm} = \boldsymbol{k}_{Cm}\,\delta_m \tag{5-3}$$

Mass transfer coefficients for different chemicals can be experimentally obtained using a two-chamber system, with both chambers of the same volume and completely mixed, with the membrane held between those two chambers. For a chemical such as oxygen (*i.e.*, the catholyte) where the concentration in one chamber is essentially constant, a mass balance yields

$$v\frac{dc_{An}}{dt} = J_m A_m = \boldsymbol{k}_{Cm}\,A_m(c_{An} - c_{Cat}) \tag{5-4}$$

where v is the liquid volume of a chamber, A_m is the membrane cross sectional area (*Kim et al.* 2007b). This mass balance is based on the lack of a chemical reaction which would consume the chemical. We assume that c_{Cat} is constant (for oxygen, its saturation concentration) at all times, and initially (t = 0) there is no chemical present in the anode chamber. Integrating and solving with these boundary conditions, we obtain an expression for the mass transfer coefficient as a function of the chemical concentration in the cathode chamber over time, or

$$\boldsymbol{k}_{Cm} = -\frac{v}{A_m t} \ln \left[\frac{(c_{Cat} - c_{An}(t))}{c_{Cat}} \right] \tag{5-5}$$

We can easily calculate the mass transfer coefficient from a plot of $-\ln c^*$ versus t, which produces a straight line, as

$$-\ln c^* = -\frac{k_{Cm} A_m}{v} t \tag{5-6}$$

where $c^* = [c_{Cat} - c_{An}(t)]/c_{Cat}$, and $\boldsymbol{k}_{\mathrm{Cm}}$ is obtained from the slope of the line based on known values of v and A_m. When monitoring dissolved oxygen it is important to use a non-consumptive DO probe (*e.g.*, Foxy-21G, Ocean Optics Inc., Fl) as the consumption of DO would affect the mass balance. The diffusion coefficient is then calculated based on the thickness of the membrane using eq. 5-3.

Example 5.2

A two-chamber reactor is used to measure the oxygen mass transfer coefficient through Nafion 117, where the membrane area is 3.5 cm^2, the bottle liquid volume is 320 mL, and the dissolved oxygen saturation concentration at ambient temperature (25°C) and pressure is 8.1 mg/L. Using the data below, calculate the mass transfer coefficient.

Time (h)	0	12	24	48	60	72
DO (mg/L)	8.1	7.6	7.0	6.0	5.5	5.1

To determine the mass transfer coefficient, we need to calculate values of c* versus time. For example, at t = 12 h, c^* is

$$c^* = [8.1 - 7.6]/8.1 = 0.0617$$

A plot of c^* versus t (with the line forced through the origin) produces a slope of $m = 0.0053$ h^{-1}, as shown in the plot below. We therefore calculate the mass transfer coefficient using eq. 5-5 as

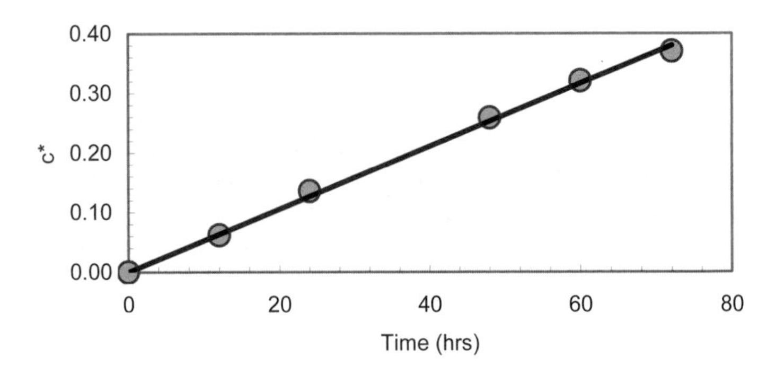

$$k_{Cm} = m\frac{v}{A_m} = (\frac{0.0053}{\text{h}})\frac{(0.320\,\text{L})}{(3.5\,\text{cm}^2)}\frac{1000\,\text{cm}^3}{\text{L}}\frac{1\,\text{h}}{3600\,\text{s}} = 1.35\times10^{-4}\,\frac{\text{cm}}{\text{s}}$$

This result compares well with the value given in Table 5.2.

For chemicals that cannot be maintained at a constant concentration in one of the chambers, the calculation is slightly different. For example, if acetate is placed into the cathode chamber at an initial concentration of $c_{Cat,0}$, the mass transfer coefficient is

$$k_{Cm} = -\frac{v}{A_m t}\ln\left[\frac{(c_{Cat,0} - 2c_{An}(t))}{c_{Cat,0}}\right] \tag{5-7}$$

The difference in this solution is that the concentration in both chambers changes over time, but the concentration of the chemical is measured only in the anode chamber. Note that either chamber can be used for these calculations as there is no chemical reaction term, but they have been written in terms of the cathode chamber for convenience.

Using the above mass balance equations, Kim *et al.* (2007b) were able to calculate the mass transfer coefficients for oxygen (k_{Om}) and acetate (k_{Am}) for several different membranes (Table 5.2). We see that Nafion is the most permeable to oxygen, with all three UF membranes much less permeable to oxygen. In contrast, the open pore size of the UF membranes made them quite permeable to acetate compared to the other membranes. The CEM membrane had the lowest permeability to acetate.

Table 5.2 Mass transfer coefficients and diffusivities of oxygen and acetate measured for cation (CEM), anion (AEM) and ultrafiltration (UF) membranes used in MFC tests (CEM = CMI-7000 and AEM = AMI-7001)

Membrane property	CEM (Nafion)	CEM (CMI)	AEM (AMI)	UF-0.5K	UF-1K	UF-3K
Thickness (cm)	0.019	0.046	0.046	0.0265	0.0265	0.0265
k_{Om} ($\times10^{-4}$ cm/s)	1.3	0.94	0.94	0.19	0.41	0.42
D_{Om} ($\times10^{-6}$ cm^2/s)	2.4	4.3	4.3	0.51	1.1	1.1
k_{Am} ($\times10^{-6}$ cm/s)	4.3	1.4	5.5	0.89	16	27
D_{Am} ($\times10^{-9}$ cm^2/s)	0.82	0.66	2.6	0.24	4.2	7.2

These different substrate and oxygen fluxes can result in decreased Coulombic efficiencies and potentially other problems as well. In mixed culture MFCs, however, some of these adverse problems are "self-repairing." For example, oxygen diffusion from the cathode into the anode chamber could result in elevating the redox potential in the anode chamber, potentially inhibiting activity or killing obligate anaerobic exoelectrogens. However, if other bacteria are present in the anode chamber that can grow aerobically (or are facultative anaerobes), they will scavenge the oxygen and help to maintain low redox potentials. Thus, the system in this case is self repairing in that a problem is eliminated by the bacteria naturally present in the system. We can expect that a biofilm will grow on the membrane (or cathode side facing the solution) and consume any oxygen that comes through before it can affect the DO or redox potentials in the anode chamber.

The leaking of substrate into the cathode chamber can cause problems, such as fouling or inactivation of the cathode catalyst. Mixed cultures again may provide a solution to this situation. Bacteria could grow on the membrane or cathode that could scavenge the substrate before it could reach the catalyst.

The permeability of the membrane to catholytes other than dissolved oxygen is not well investigated. In particular, the potential for ferricyanide to cross different membranes has not been investigated. Ferricyanide can function as an electron mediator (*Bond et al.* 2002; *Emde et al.* 1989), raising the possibility that leakage of this chemical into the anode chamber could increase power densities by functioning as an inadvertent exogenous mediator. Alternatively, the diffusion of ferricyanide and other chemicals into the anode chamber could trigger stress responses by bacteria that might result in release of stress-promoted chemicals that could act as endogenous mediators. Such possible situations deserve greater attention and measurements in laboratory experiments, particularly in tests where the anode chamber contents are not regularly flushed as a consequence of fed-batch or continuous-flow operations (*Rabaey et al.* 2003).

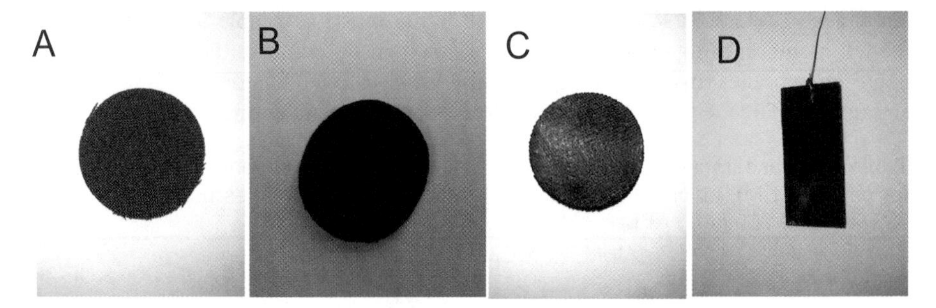

Fig. 5.5 Photographs of some cathode materials. Single-chamber cathodes: (A) plain carbon cloth; (B) carbon cloth coated with Pt catalyst on one side that typically faces the liquid; (C) carbon cloth with a diffusion layer applied that typically faces the air. (D) Square cathode used in two-chamber systems that is suspended in the water.

5.4 Cathode materials

The design of the cathode is the single greatest challenge for making an MFC a useful and scalable technology. The chemical reaction that occurs at the cathode is difficult to engineer as the electrons, protons and oxygen must all meet at a catalyst in a tri-phase reaction (solid catalyst, air, and water). The catalyst must be on a conductive surface, but it must be exposed to both water and air so that protons and electrons in these different phases can reach the same point. Oxygen can diffuse into water, but the solubility of oxygen (mole fraction basis) in water is only 4.6×10^{-6} (25°C) compared to 0.21 in air. Proton solubility in membranes or binders such as Nafion can be high producing a low pH, but the proton concentration in water is limited by pH ranges tolerated by the bacteria where are circum-neutral (*i.e.,* 10^{-7} M). Lower pHs can be used in the cathode chamber, but this can result in voltage losses across the membrane as discussed above.

The same materials that have been described above for use as the anode have also been used as cathodes. Thus, studies have used carbon paper, cloth, graphite, woven graphite, graphite granules, brushes, *etc.* The main difference when these materials are used for the cathode is that a catalyst is usually (*i.e.,* Pt for oxygen reduction) but not always needed (*i.e.,* ferricyanide). Solid phase and liquid catalysts have been used, creating a wide range of possible materials and chemicals to facilitate current generation.

5.4.1 Carbon-based cathodes

Carbon cathodes with Pt catalysts. The most commonly used material for a cathode is commercially available carbon paper pre-loaded with a Pt catalyst on one side, available from different manufacturers (*e.g.,* E-Tek, USA, 0.35 mg-Pt/cm^2). When used in the MFC, the side containing the catalyst faces the water, with the uncoated side facing air. It is also possible to purchase plain carbon cloth and apply the catalyst in the laboratory (Fig. 5.5A). With experience, you can make electrodes that will perform better than commercially manufactured electrodes with the same Pt loading. To make a Pt-catalyst cathode take commercially available Pt (10 wt% Pt/C, E-TEK) and mix it with a chemical binder such as a 5% Nafion™ liquid solution to form a paste (7 µl-binder per mg-Pt/C catalyst) (*Cheng et al.* 2006c). Apply the paste evenly to one side of the electrode and dry at room temperature for 24 hours (Nafion) (Fig. 5.5B, 5.5D). Cheng *et al.* (2006c) found that varying the Pt content from 0.1–2 mg/cm^2 did not appreciably affect power densities in air-cathode MFCs.

Catalyst binders. When the catalyst is applied to carbon it must be held there using a material that allows transfer of protons, electrons and oxygen. Nafion is therefore typically used due to its high proton conductivity and oxygen permeability. Other materials such as polytetrafluoroethylene suspension (PTFE) can also be used. Cheng *et al.* (2006c) prepared Pt-based cathodes as described above, except a 2% solution of PTFE was used instead of Nafion, and the electrode was dried at 350°C for 0.5 hours. They found that Nafion produced slightly greater (more positive) potentials in electrochemical (abiotic) tests than those prepared using PTFE at a Pt loading of 0.5 mg/cm^2 (for example, 12% more at at a current density of 1 mA/cm^2). Air-cathode MFCs with Nafion produced more power (480 ± 20 mW/m^2) than those with PTFE (360 ± 10 mW/m^2). However, the maximum power densities declined more with Nafion (19%) than with PTFE (9%) in tests conducted over 34 cycles of fed batch operation. While electrochemical tests showed little effect of Pt loading on cathode potential, MFC tests with the Nafion binder showed that power was reduced by ~19% using a 0.1 mg/cm^2, compared to that with 0.5 mg/cm^2.

Diffusion layers. When air-cathode MFCs do not contain a CEM, Coulombic efficiency (C_E) is reduced due to the high oxygen flux through the cathode. In addition, there can be substantial water loss through the air-facing side. In some reactors that can result in the appearance of a gas headspace which could be composed of carbon dioxide, methane, nitrogen, and oxygen depending on operational conditions. Obviously, the occurrence of an air phase in the anode chamber is to be avoided as the oxygen in the air may not only reduce the C_E but also affect power generation if the redox potentials become too high in the anode chamber. To avoid these problems of water loss and low C_E, Cheng *et al.* (2006a) added a hydrophobic coating onto the air-facing side of the cathode.

While a Nafion CEM can also serve the same purpose, it is used on the water-side of the cathode, not the air-side. The amount of material that needed to be applied to the cathode was investigated by applying different loadings, with each loading called an additional diffusion layer (DL). They first applied a carbon base layer (0-DL) to the cathode using a mixture of carbon powder (Vulcan XC-72) and 30 wt% PTFE solution (20 μl per mg of carbon power) onto one side of the carbon cloth. This base layer (2.5 mg/cm^2) was air-dried at room temperature for 2 h and then heated for 0.5 h at 370°C. Additional layers (2 to 8 DLs) were made by brushing on a PTFE solution (60 wt %) onto the base layer, each time drying and heating (370°C for 10 min) (Fig. 5.5C). Following this procedure, they applied the Pt catalyst (0.5 mg/cm^2) to the other side of the cathode.

In MFC tests, they found power production was highest with the addition of 4 DLs (766 mW/m^2). Power without a DL was 538 ± 6 mW/m^2, and if more than 4 DLs were applied power decreased likely as a result of insufficient oxygen transfer through the cathode. The measured oxygen mass transfer coefficients for a cathode lacking a DL was 3.3×10^{-3} cm/s and increased slightly (18%) with the base layer coating to 3.9×10^{-3} cm/s, likely as a consequence of the heat treatment while applying the coating. With 4 DLs, the mass transfer coefficient decreased to 2.3×10^{-3} cm/s, decreasing further to 0.6×10^{-3} cm/s for 8 DLs. The CEs increased with the number of DLs, ranging from 13–20% in the absence of the DL to 20–27 % with four DLs. C_E increased with successive DLs, reaching a maximum of 32% for 8 DLs due to reduced oxygen diffusion rates into the anode chamber.

Carbon cathodes with non-Pt catalysts. Park and Zeikus (2002)first experimented with non-precious-metal, carbon-based air cathodes in MFCs. They made ferric (Fe^{3+}) cathodes by forming plates out of ferric sulfate (3% w/w), fine graphite (60%), kaolin (36%, as a binder), and nickel chloride (1%) and baking at 1100°C for 12 h under N_2 gas. These iron cathodes produced up to 3.8 times as much power as plain woven graphite cathodes, but they were not compared to Pt-based cathodes of similar dimensions (*Park and Zeikus* 2003).

Cathode performance equal to that of Pt-based carbon cathodes has now been achieved using transition-metal carbon cathodes, thus eliminating the need for precious metals in MFCs. Zhao *et al.* (2005) showed in electrochemical tests that two different transition metal catalysts, iron(II) phthalocyanine (FePc) or cobalt tetramethoxyphenylporphyrin (CoTMPP), could produce power at levels comparable to or better than those achieved with Pt-based cathodes at current densities above 0.2 mA/cm^2. At higher current densities, CoTMPP performed slightly better than FePc. Both materials performed better at low pH, but noted that bacterial suspensions could not be used at those low pH values. In tests using biohydrogen fuel cells (bHFCs) (*e.g.*, with substrate fermentation by *E. coli* producing H_2), maximum power volumetric densities were 14 W/m^3 with either catalyst, but power output was not compared to a Pt-catalyst.

The use of CoTMPP as a catalyst was also examined in a series of MFC tests by Cheng *et al.* (2006c) using air-cathodes. The maximum power density with CoTMPP (0.6 mg/cm^2) was 369 ± 8 mW/m^2, which was found to be only 12% lower than that with 0.5 mg-Pt/cm^2. However, this power output was larger than that obtained using 0.1 mg-Pt/cm^2 on the cathode (340 ± 20 mW/m^2) under the same operational conditions. The C_E values ranged from 7.9–16.3% and were similar to those with the Pt catalyst. These power densities are substantially larger than those obtained in the absence of a catalyst on the cathode which averaged 93 ± 13 mW/m^2.

Plain carbon cathodes. The efficiency of a catalyst is often assessed by comparing current or power densities to those with plain carbon electrodes of the same surface area. Oxygen reduction still proceeds in the absence of the catalyst, but the rate is reduced. In general, current and power are reduced by a factor of 10 or more with plain carbon materials. However, if the cathode surface area is substantially increased, it is possible to achieve much higher power densities. Reimers *et al.* (2006) tested a sediment MFC that had 1-m long carbon brush cathodes. They reached power densities of 34 mW/m^2 by positioning the MFC over a deep ocean cold seep in Monterey Canyon in California. Power decreased over time due to sulfide build up on the anode with no indication of

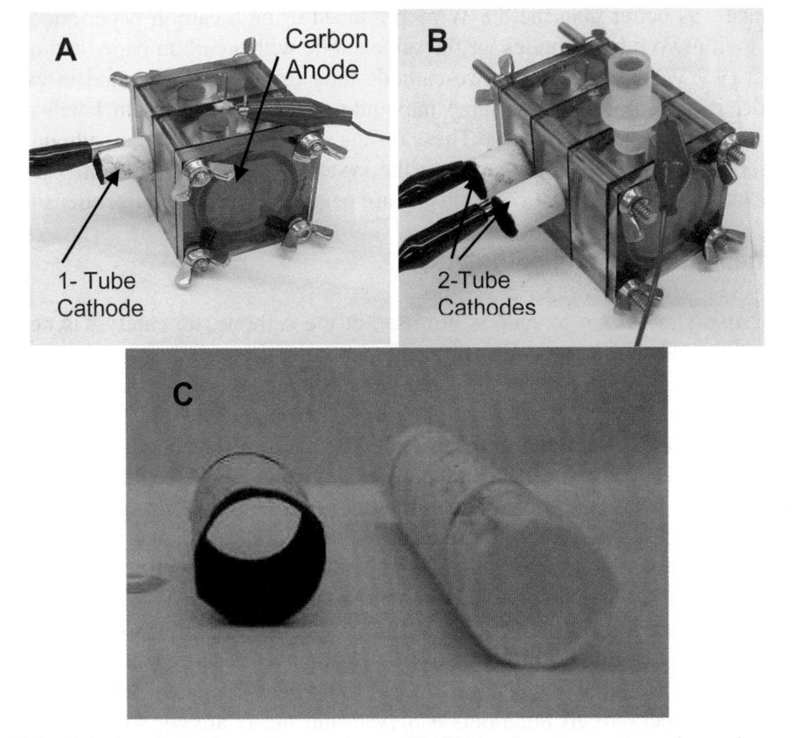

Fig. 5.6 Tubular cathode cube reactors with (A) carbon paper anode and one tube cathode; and (B) carbon paper anode and two cathode tubes. [Reprinted from *Zuo et al.* (2007) with permission from the American Chemical Society.] (C) Close-up of a treated and untreated tubular ultrafiltration membrane used to make the conductive cathode.

reduced cathode performance. The use of carbon brush cathodes was first tested by Hasvold *et al.* (1997) with magnesium alloy anodes (*i.e.*, not in an MFC application). These cathodes were used in systems that produced 650 kWh (*Hasvold et al.* 1997) and 600 W at 30 V (*Hasvold et al.* 1999), attesting to their ability to sustain high levels of power generation.

Tubular carbon-coated cathodes. MFCs require high surface areas and porosities typical of wastewater reactors. One new approach to wastewater treatment has been to use tubular ultrafiltration membranes, providing high surface areas for filtering the treated water (180 to 6800 m^2/m^3). Based on that idea of high surface areas provided by these membranes, Zuo *et al.* (2007) developed a tubular cathode by applying a conductive graphite paint material to a hydrophilic tubular ultrafiltration membrane (polysulfone membrane on a composite polyester carrier) that had an inner diameter of 14.4 mm (B0125, X-FLOW) and wall thickness of 0.6 mm (Fig. 5.6). The tubes were coated on the air-facing side with a proprietary graphite paint (ELC E34 Semi-Colloidal, Superior Graphite Co.) to make the tube electrically conductive. They used a non-precious metal CoTMPP catalyst applied to the paint (*Cheng et al.* 2006c). These tubular cathodes were tested in cube-type MFCs, with the tubes cut to a lengths of 3 cm. Using a graphite brush anode ($A_{an,s}$ = 7700 m^2/m^3) with two tube cathodes placed inside the reactor ($A_{cat,s}$ = 93 m^2/m^3), the MFC produced 18 W/m^3 with a C_E of 70–74%. This performance was better than the 8.8 W/m^3 produced using a carbon paper anode ($A_{an,s}$ = 25 m^2/m^3) with two tube-cathodes, or the same anode with a carbon paper cathode with a Pt catalyst (9.9 W/m^3). When the two-cathode tube reactor was operated in continuous flow mode, it also produced the same maximum power density as in batch mode (18 W/m^3) with a COD removal of 38%. These results show that the tubular cathode design is a very promising approach for scaling up the system, but membranes with less internal resistance and greater electrical conductivity are needed to increase power densities.

5.4.2 Other cathodes and catholytes

Aqueous catholytes. When oxygen is not used at the cathode, no catalyst is needed and therefore plain carbon cathodes can be used. Several different aqueous catholytes have now been tested, but the most common is ferricyanide, or hexacyanoferrate which is reduced according to $Fe(CN)_6^{3-} + e^- \rightarrow Fe(CN)_6^{4-}$. Permanganate and iron have also been used as catholytes, as discussed in Chapter 3. The main disadvantage of catholytes like ferricyanide is that they must be chemically regenerated or replaced. A 100 mM concentration of potassium hexacyanoferrate is commonly (*Park and Zeikus* 2000; *Rabaey et al.* 2003) prepared in the same buffer solution used for the anode. Sometimes these solutions are aerated (*Rabaey et al.* 2003), although the oxygen reduction rate in the absence of a catalyst will be quite low relative to the reactivity of the ferricyanide solution. The presence of dissolved oxygen means that oxygen can still diffuse into the anode chamber. Some of the highest power densities yet recorded have been achieved with ferricyanide as catholytes (*Rabaey et al.* 2004; *Rabaey et al.* 2003), as a result of both the higher open-circuit potentials achieved and the relatively small overpotentials compared to oxygen. In side-by-side tests with two-chamber MFCs, Oh *et al.* (2004) found an increase of only 1.5–1.8 times using ferricyanide versus dissolved oxygen. However, internal resistances in these reactors were high and therefore power was likely limited by other factors, such as electrode spacing and a small CEM.

Pt and Pt-coated metals. Pt electrodes are not practical for large-scale applications due to their prohibitive cost, but they provide useful benchmarks on the performance of the system. While Pt or platinized Pt electrodes can be used in MFC tests, the formation of a oxide layer (PtO) on the Pt surface can reduce electrode activity over time (2003).

Metals other than Pt. Cathodes made of several different metals have been examined for use in various types of fuel cells, although not all of them can be considered to be MFCs. For example, the system used by Shantaram *et al.* (2005) was briefly described in Chapter 3. This is not a true MFC as the system used a sacrificial metal anode (Mg). In other tests they used glucose and a culture of *Klebsiella pneumoniae* and a mediator (2-hydroxy-1,4-naphthoquinone, HNQ) in the anode chamber, which also means that they did not test a true mediator-less MFC (*Rhoads et al.* 2005). In both cases, the cathode was

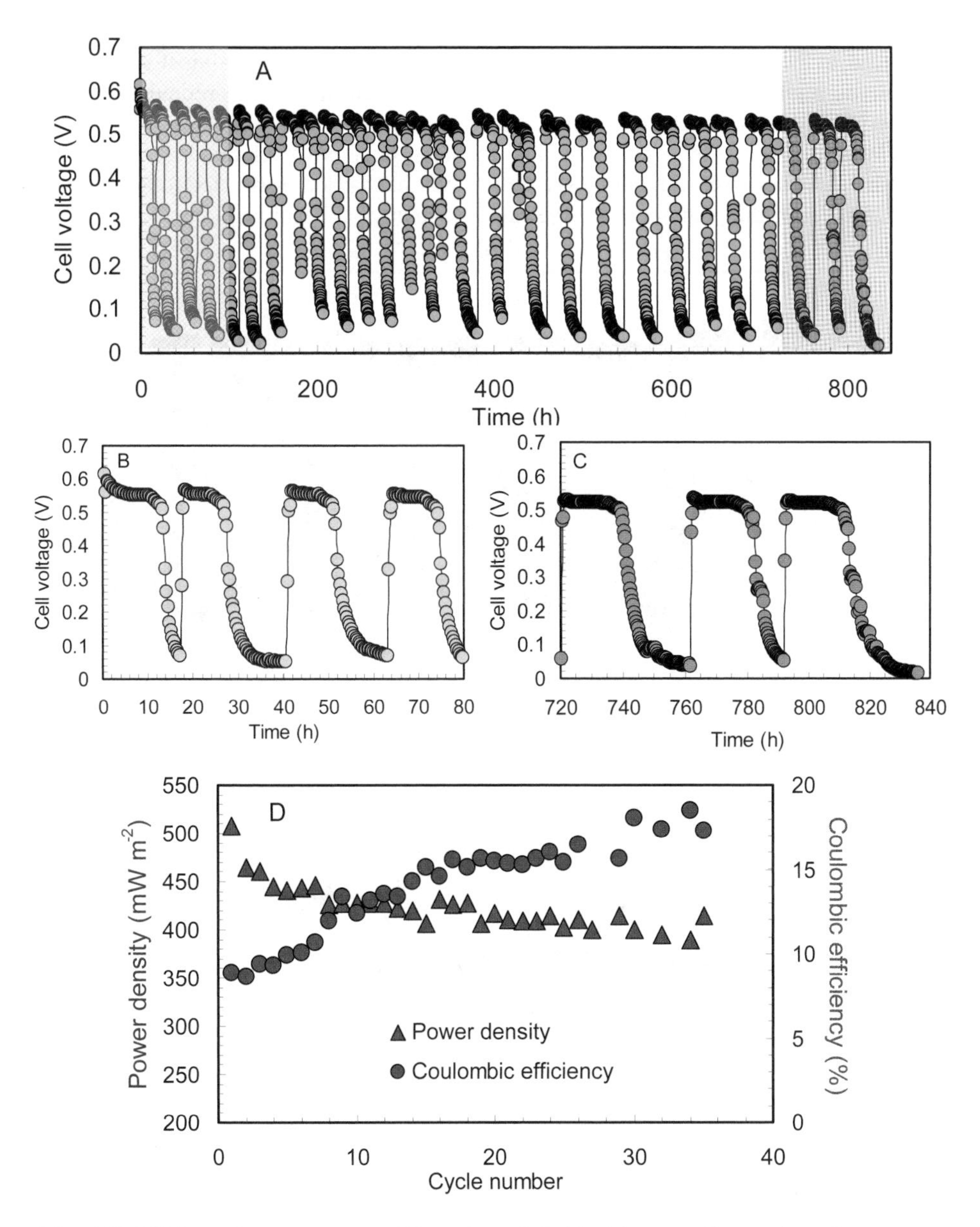

Fig. 5.7 MFC performance over 35 cycles of fed-batch operation using glucose as substrate with a Nafion binder and a Pt catalyst (0.5 mg/cm^2). (A) Cell voltage; (B) expanded view of for the first four cycles and (C) last three cycles; (D) power density and Coulombic efficiency as function of number of cycles. [Reprinted from *Cheng et al.* (2006c) with permission from the American Chemical Society.]

made of solid MnO_2. This could be considered a type of biocathode as the authors added a culture of *Leptothrix discophora* SP-6 to oxidize the reduced Mn^{2+} to MnO_2. To allow for Mn oxidation by the bacteria, they aerated the cathode chamber to ensure adequate concentrations of dissolved oxygen. They found that the maximum power density produced in the system using glucose was 127 mW/m^2, which was almost two orders of magnitude higher than achieved with dissolved oxygen alone with reticulated vitreous carbon (RVC) lacking a catalyst.

Hasvold *et al.* (1997) reported that stainless steel brush electrodes were used as cathodes in an undersea vehicle power source, but that carbon brush electrodes performed better. However, they did not test these cathodes in any MFC applications.

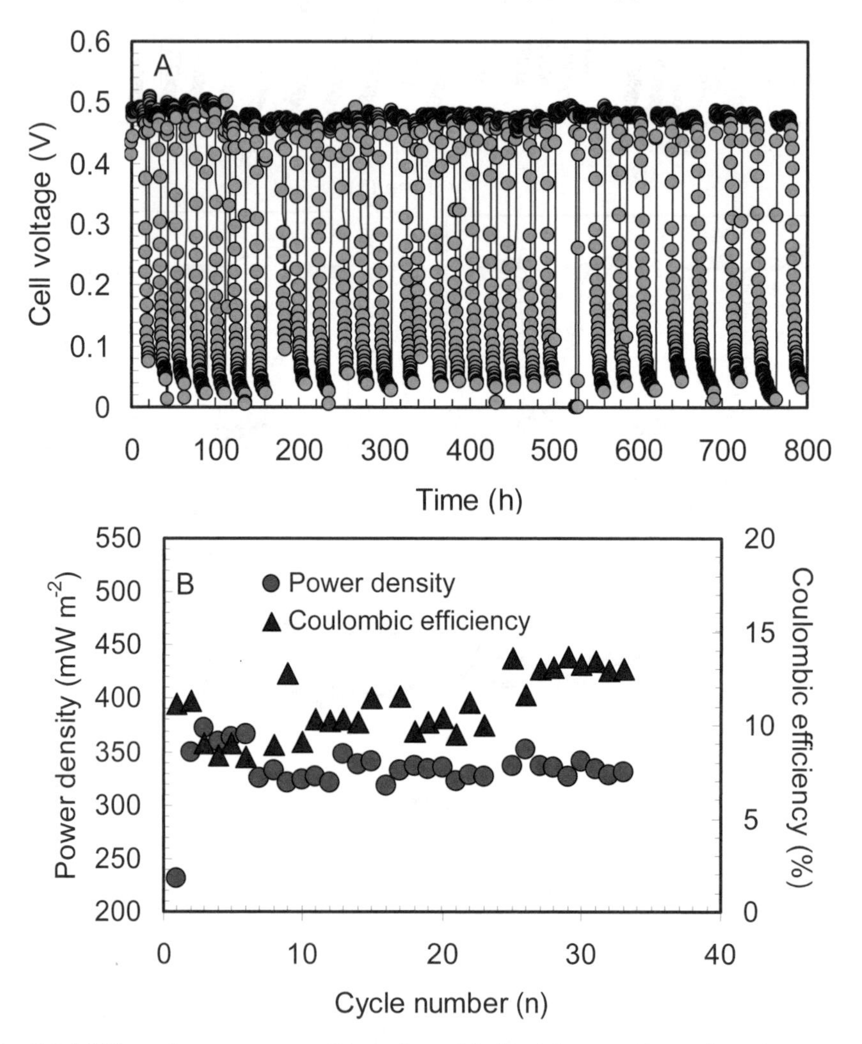

Fig. 5.8 MFC performance over 34 cycles of fed-batch operation using glucose as substrate with PTFE as a binder and a Pt catalyst (0.5 mg/cm^2). (A) Cell voltage; (B) power density and Coulombic efficiency as function of number of cycles. [Reprinted from *Cheng et al.* (2006c), with permission from the American Chemical Society.]

Biocathodes. One relatively new area of development is the concept of using bacteria as the catalysts on the cathode. When stainless steel cathodes (100-cm^2 plates) containing a seawater biofilm were used for a HFC, power was increased by a factor of 30 compared to the same system after cleaning the biofilm on the cathode (*Bergel et al.* 2005). As noted above, systems that used bacteria to re-oxidize Mn also can be considered to be biocathodes. In tests with an MFC system using aqueous cathode based on conversion of ferric to ferrous iron, it was suggested that the iron could be re-oxidized through bacterial assisted catalysis (*ter Heijne et al.* 2006). In other non-MFC based tests (*i.e.,* without bacteria at the anode), it has been shown that nitrate could be reduced to nitrite by *Geobacter metallireducens* when the anode was poised at a high potential. Other electrochemical tests have been performed to show bacteria reduction of different compounds, but these systems also required the voltage of the system to be augmented with external power sources as summarized in a recent review.

The first real breakthrough in biocathodes was achieved by Clauwaert *et al.* (2007), who demonstrated that nitrate could be used as the catholyte in an MFC. Acetate was oxidized at the anode, and nitrate was completely converted to nitrogen gas, producing a maximum of 4 W/m^3 (cathode chamber volume), cell voltage of 0.214 V, and a current density of 35 A/m^3. The reactor was a tubular design with the cathode placed inside the anode in a concentric cylinder orientation, with the two chambers filled with graphite granules and separated by a CEM. Both the dentrification rate and power production were limited by cathode performance, which did not result in denitrification activity below a cathode potential of 0V (vs. SHE).

5.5 Long-term stability of different materials

The long-term stability of the electrode materials and membranes, particularly due to the biofilm formation on these surfaces, is not well studied. Several researchers have indicated that laboratory systems have run for months to years. Detailed information on the performance of these systems is not published, however, so it is difficult to determine if there were operational problems or how upsets (lack of feeding, temperature and pH variations) affected power generation. In general, it seems that power generation is stable if reactors are continuously fed, or if there are not long lapses (several days or more) in re-feeding reactors operated in fed-batch mode. The most serious effects on power generation seem to be consistent with adverse effects caused to any biological system, *i.e.*, excessive pH or ionic strength changes. One reactor in our laboratory continuously fed sodium acetate, for example, failed after successive feedings due to a buildup of sodium (*i.e.,* a high salt concentration). In other tests it has been found if the anode dries out, subsequent power generation by the system after re-hydration and feeding will be severely impaired although long term recovery of the system was not studied. However, long-term operation of an MFC can lead to shifts in the microbial community composition that improves power generation. Aelterman *et al.* (2006) found that over time, power increased by a factor of 3.8, and internal resistance decreased from 6.5 to 3.9 Ω, accompanied by a shift in the microbial community.

Stability of power generation over time with a fed-batch MFC. In one study detailed information was provided on power production and the Coulombic efficiency (C_E) over repeated feeding cycles in an MFC with Pt catalysts bound using two different binder materials: Nafion or polytetrafluoroethylene (PTFE) (*Cheng et al.* 2006c). MFC fed batch tests were conducted for 35 cycles, spanning 31 days, with glucose as the substrate. In the MFC using the Nafion binder, the voltage varied between 0.5 V and 0.6 V but decreased slightly over time (Fig. 5.7). While the maximum power density decreased from 480 ± 20 mW m^{-2} (± S.D. based on the first three cycles) to 400 ± 10 mW m^{-2} (last three cycles), the C_E increased over the same period from 8.9 ± 0.4% to 18.6 ± 0.5%. This suggests that both proton conduction and oxygen diffusion were impaired over time. The development of a biofilm on the cathode, for example, could reduce the effective diffusivity of protons from the anode to cathode. This biofilm, however, might also cause a reduction of substrate and oxygen diffusivities near the cathode, reducing overall substrate losses due to bacterial aerobic growth sustained by oxygen diffusion into the reactor.

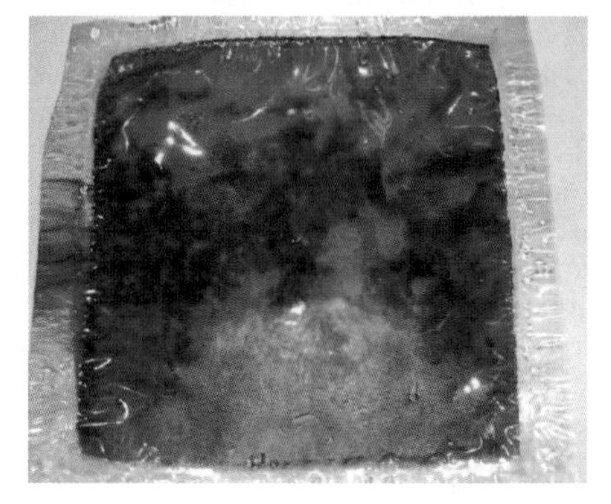

Fig. 5.9 Fouling of a Nafion 117 membrane due to a pH rise in the cathode chamber, causing ferric iron precipitation. [Reprinted from ter *Heijne et al.* (2006), in supplemental information, with permission from the American Chemical Society.]

When PTFE was used as the binder instead of Nafion, the same general pattern was observed with decreased maximum power densities and increased C_E over time (Fig. 5.8). The maximum voltages and power densities with PTFE were not as high as those with Nafion, but the changes with time were reduced as well. The maximum power decreased by only 9%, or from 360 ± 10 mW/m^2 (cycles 2–4) to 331 ± 3 mW/m^2 (last three cycles) compared to a 19% decrease with Nafion (Fig. 5.8B). The C_E increased from 9.5 ± 1.5 % to 13.1 ± 0.3 %. Thus, the behavior of this system followed the same general trend as that with Nafion, but the magnitudes of the changes differed suggesting that the binder material had some effect on the overall process.

Membrane fouling. For MFC systems that contain a membrane, fouling of the membrane can limit chemical diffusivities through it. This can reduce proton transport or charge transfer, as well as oxygen or substrate diffusion between the chambers. In one study it was noted that when using a ferric iron catholyte that the CEM quickly became fouled due to iron precipitation on the membrane (Fig. 5.9) (*ter Heijne et al.* 2006). This problem was solved by replacing the CEM with a bipolar membrane. In other laboratory studies black deposits are sometimes observed on Nafion membranes, and these appear to develop at the same time sulfide odors appear. However, the consequences of such deposits and odors have not been studied. The long term effects of membrane fouling will need to be investigated, but this will likely need to wait for the development and testing of larger-scale systems.

CHAPTER 6

Architecture

6.1 General requirements

A large range of materials have been used in MFCs, as summarized in the previous chapter. How these materials are packed, arranged, and used in the final system design—what is referred to here as the reactor architecture—ultimately dictates how the system will perform in terms of power output, Coulombic efficiency, stability, and longevity. Many researchers have chosen to use air-cathodes, as these types of electrodes will ultimately be the type of cathodes used in larger systems. The first air-cathode MFC was reported by Sell *et al.* (1989), but this study was apparently unknown to many MFC researchers as this paper was not cited in early air-cathode studies. Other researchers have sought to explore the maximum power densities possible with bacteria by using optimized cathodes based on chemical reductions with chemicals such as ferricyanide or permanganate. Sometimes the power produced is not as interesting as the bacteria that grow in the reactor. For these studies, two-chamber systems having high internal resistance are suitable to determine if power generation is possible with a particular substrate or what microbial communities might evolve in the system as a consequence of the inoculum, but they don't distinguish whether different microorganisms or substrates might improve performance. For laboratory studies, we can therefore imagine a wide range of useful architectures depending on the goals of the research.

Practical applications of MFCs will require that we develop a design that will not only produce high power and Coulombic efficiencies, but one that is also economical to mass produce based on the materials being affordable and the manufacturing process being practical to implement on a large scale. While the reactor designs that will ultimately prove to meet these requirements of power, efficiency, stability, and longevity are still being developed, we now know that scalable and economical systems can be developed using graphite fiber brush electrodes and tubular cathodes immersed together in a tank. However, such a reactor has yet to be built at pilot or large scale to date. Thus, the final design and the materials that will ultimately be used in a large-scale system remain unproven at this time.

This chapter is intended to give the reader an appreciation of the various factors that can affect power production in an MFC and to see all the marvelous ways that different materials have been used in MFC designs to test the factors that can affect power production. The most extensive review of these systems is done in the next section, focusing on the various reactors that have been tested that use air cathodes. This includes a description of a scalable system based on a two recent papers and a patent filed by Penn State. In successive sections, other types of systems are explored based on different catholytes, packing materials, operation modes (stacked cells) and energy input (fermentation, photobiological, and metal cathodes). New innovations will no doubt continue to modify our views on the "perfect" MFC system, with improvements certain to be made in the coming years. There is little doubt that practical reactors will soon emerge based on electricity and hydrogen generation (see Chapter 8), and with increased funding for this area we can envision systems that not only provide power to a community, but that ensure the energy sustainability of our water infrastructure. A vision for the implementation of a scaled up system for wastewater treatment is described in Chapter 9.

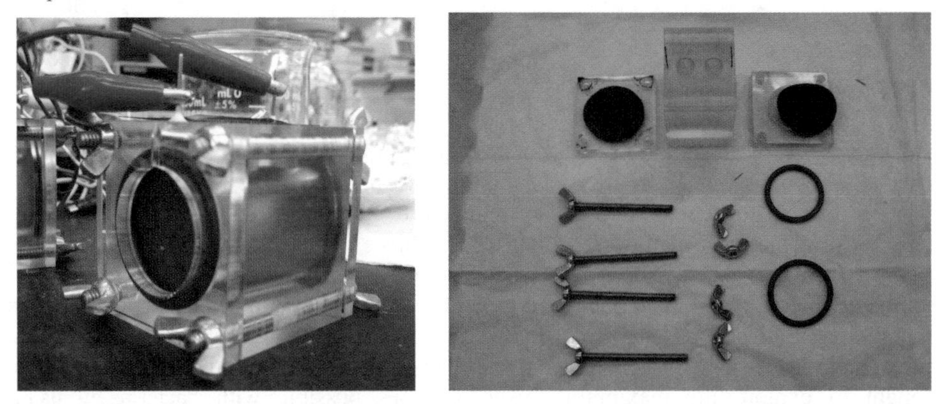

Fig. 6.1 Single chamber air-cathode cube system based on the design of Liu and Logan (2004) shown (A) assembled and (B) unassembled with all parts except the stoppers for the top filling ports.

6.2 Air-cathode MFCs

Cube reactors. A useful and simple design for examining factors that affect power production in MFCs was an air-cathode reactor developed at Penn State (*Liu and Logan* 2004). This the single-chamber, air-cathode cube reactor has been has been used and reported on in over a dozen publications (Fig. 6.1). The simple design of the reactor allows examination of a variety of factors on power production. The cube reactor consists of a single 4-cm block of Acrylic or Lexan (a material that can be autoclaved) drilled through producing a 3-cm-diameter chamber. The empty bed volume is 28 mL, and when the two electrodes are placed on opposite ends the surface area per volume of the reactor (based on the anode projected surface area) is 25 m^2/m^3. Two openings are made on the top to allow the reactor to be easily drained and filled. These are sealed with thick stoppers to prevent oxygen from entering the reactor when the reactor is operated. The electrodes sit in the ends that are cut to be slightly recessed, a round rubber gasket placed over the electrode, and then the flat plate ends are attached on top of the gasket to form a water tight seal. In the first tests the anode was made of Toray carbon paper (without wet proofing; E-Tek, USA; no catalyst), although in later tests other materials were used. The cathode is typically carbon cloth containing 0.5 mg/cm^2 of Pt catalyst (E-Tek, USA) facing the water-side of the reactor. In the first system developed this was covered by a cation exchange membrane (CEM) of Nafion™ 117 (Dupont, USA). The CEM was hot pressed directly onto the cathode by heating it to 140°C at 1780 kPa for 3 min. In other tests rigid carbon paper was used for the cathode that contained 0.35 mg/cm^2 of Pt (E-Tek, USA). Two pieces of platinum (or stainless steel) wire are inserted through drilled holes that line up with the bottom of the recessed opening so that good contact is made when the electrode is inserted and pressed onto the wire. Four screw bars are then used to compress the end plates onto the ends, holding the reactor together. The circuit is completed with an external resistor, usually 500 or 1000 Ω.

The power produced with glucose in the original tests was 494 ± 21 mW/m^2 (12.5 ± 0.5 W/m^3) in the absence of the CEM, and 262 ± 10 mW/m^2 (6.6 ± 0.3 W/m^3) with the CEM (*Liu and Logan* 2004). With domestic wastewater the MFC without the CEM generated 146 ± 8 mW/m^2 (3.7 ± 0.2 W/m^3), versus 28 ± 3 mW/m^2 with the CEM. Coulombic efficiencies (C_E) for glucose were 40–55% (CEM) or 9–12% (no CEM), and for wastewater were 28% (CEM) and 20% (no CEM). The CEM was found to adversely affect the cathode potential, producing 0.425 V without the CEM and producing 0.226 V with the CEM in tests with glucose. Further tests were conducted with acetate and butyrate in the same reactor in the absence of the CEM. Acetate produced approximately the same amount of power (506 mW/m^2 or 12.7 W/m^3) as glucose, but butyrate produced less power (305 mW/m^2) (*Liu et al.* 2005b). C_Es were also higher with acetate (9.9% to 31.4%) than butyrate (15% to 7.8%).

Fig. 6.2 Cube reactor designed by Cheng *et al.* (2006b) to have removable sections so that the anode can be placed at distances of 4 cm, 2 cm and 1 cm from the cathode.

Subsequent tests with the cube reactor were conducted to find ways to increase power generation. The same reactor was used, but this time the anode was pierced with a few holes to allow fluid to move from one side to the other when the anode was moved closer to the cathode. Decreasing the electrode spacing from 4 to 2 cm increased power output from 720 mW/m^2 to 1210 mW/m^2 with acetate as a substrate due to a reduction in internal resistance of the reactor (*Liu et al.* 2005a). Also, increasing the solution ionic strength from 100 to 400 mM using NaCl increased power output from 720 mW/m^2 to 1330 mW/m^2. Using a cathode made by the authors instead of a commercial cathode increased power by 68%, and decreasing the temperature from 32°C to 20°C reduced power output by only 9%. C_E reached a maximum of 61.4%, and the maximum energy recovery was 15.1%. A further decrease in the electrode spacing to 1 cm (Fig. 6.2) reduced power (from 811 mW/m^2 to 423 mW/m^2; with glucose), despite a reduction in internal resistance (*Cheng et al.* 2006b). However, it was shown that with a 1-cm electrode spacing power could be increased to 1540 mW/m^2 (51 W/m^3; C_E = 60%) when the reactor was operated in continuous flow mode with the flow directed through the anode towards the cathode. Presumably, this advective flow helped to reduce the effect of oxygen diffusion from the cathode to the anode. With domestic wastewater (255 ± 10 mg-COD/L), the maximum power produced by the cube reactor with advective flow through the anode was 464 mW/m^2 (15.5 W/m^3; C_E = 27 %) (*Cheng et al.* 2006b).

Other substrates have been used in the cube reactor, under differing conditions of substrate concentration, ionic strength, and temperature. These substrates include proteins like bovine serum albumin, a meat packing wastewater (*Heilmann and Logan* 2006), propionate (*Oh and Logan* 2005), ethanol (*Kim et al.* 2007c), corn stover hydrolysates (liquefied corn stover produced by a steam explosion process) (*Zuo et al.* 2005), food processing wastewaters (*Oh and Logan* 2005), and animal (swine) wastewater (*Min et al.* 2005b). In general, power production has been equal to or lower than that achieved with acetate and glucose. There currently is no way to predict the level of power that can be achieved with a particular substrate in an MFC without direct tests.

Materials and solution chemistry have been show to substantially affect power production. While these factors have been described in the previous chapter, they are briefly summarized here to provide context for results with cube reactors. For example, using a phosphate buffer to increase solution ionic strength in an MFC with a 2-cm electrode spacing increased power to 1640 mW/m (96 W/m^3) (*Cheng and Logan* 2007). Treating the anode with a high-temperature ammonia gas procedure further increased power to 1970 mW/m^2 (115 W/m^3) (*Cheng and Logan* 2007). Replacing the flat carbon paper or carbon cloth anode with an ammonia-treated graphite brush electrode (Fig. 5.3C) in an MFC 4 cm in length (not a 4-cm spacing as the brush electrode spanned the anode chamber) increased power to 2400 mW/m^2 (73 W/m^3; CE = 60%) (*Logan et al.* 2007).

Fig. 6.3 Cube reactor containing a brush anode and two tubular cathodes used by Zuo *et al.* (2007). A brush and tube cathode are also shown next to the reactor.

The lower volumetric power density than that produced with the flat anode reflects the larger volume of reactor used to accommodate the brush electrode.

Various binders for Pt, and different cathode catalysts have been explored in order to avoid the need for using a precious metal (Pt) on the cathode. Nafion performs better as a Pt catalyst binder than PTFE (*Cheng et al.* 2006c). It was demonstrated by Cheng *et al.* (2006c) that Pt loadings could be decreased from 0.5 to 0.1 mg/cm^2 with no substantial effect on maximum power densities in cube reactors. In addition, the Pt could be replaced by a cobalt-based material (Co-tetramethyl phenylporphyrin; CoTMPP) with only a slightly reduced (12%) effect on power. Additional tests showed that a high Co-TMPP loading produced similar power densities as Pt (*Cheng et al.* 2006c; *Cheng and Logan* 2007; *Yu et al.* 2007), consistent with findings by others using a different type of reactor (*Rosenbaum et al.* 2006).

A recent change in the cube reactor is the use of a tubular cathode instead of a flat cathode placed at the opposite end of the chamber from the anode (Fig. 5.6C). A single tube cathode consisted of a 3-cm-long hydrophilic ultrafiltration (UF) tubular membrane (polysulfone on a composite polyester carrier) with an inner diameter of 14.4 mm (B0125, X-FLOW). The tube was coated with a proprietary conductive graphite paint two times (ELC E34 Semi-Colloidal, Superior Graphite Co.) and was then coated with a cobalt catalyst (CoTMPP). A two-tube cathode reactor with a brush anode is shown in Fig. 6.3. Using glucose as a substrate and also using a graphite brush anode ($A_{An,s}$ = 7700 m^2/m^3) with two tube cathodes placed inside the reactor (A_{cat} = 27 cm^2, $A_{cat,s}$ = 93 m^2/m^3), the power produced by this MFC was 18 W/m^3 with a C_E = 70–74 %. Further increases in the surface area of the tube cathodes to 54 cm^2 (120 m^2/m^3) increased the total power output (from 0.51 to 0.83 mW), but the increase in volume resulted in a constant volumetric power density (~18 W/m^3). Power produced with the typical flat carbon paper anode (A_{An} = 7 cm^2, $A_{An,s}$ = 25 m^2/m^3) and two cathodes with CoTMPP was 403 mW/m^2. This power was similar to that obtained with a carbon paper cathode containing a Pt catalyst (394 mW/m^2; A_{cat} = 7 cm^2, $A_{cat,s}$ = 25 m^2/m^3). These power densities are lower than previously achieved with systems lacking a CEM due to the added internal resistance of the membrane (*Kim et al.* 2007b). However, the benefit is that this configuration of cathode tubes and anode brushes provides a scalable design (*Logan* 2005). It should be possible through improvements in membrane materials to improve power generation with these tube-cathode systems.

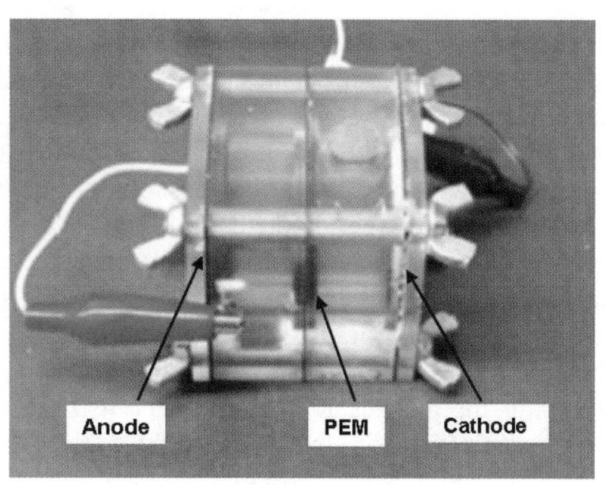

Fig. 6.4 Two-chamber, air-cathode cube type MFC. The CEM is placed in the middle of the chamber, making it possible to examine the effect of different membrane types under similar conditions and without the need to bond the membrane to the cathode. [From *Kim et al.* (2007b), reprinted with permission from the American Chemical Society.]

Two-chamber air-cathode system. The 4-cm reactor with an air cathode was adapted by Kim *et al.* (2007b) to study the effect of different membranes on internal resistance. A membrane was placed into the middle of the reactor, creating two equally sized chambers: one where the bacteria and solution were placed, and the second containing just buffered solution (Fig. 6.4). Several different cation, anion, and ultrafiltration membranes were used as described in the previous chapter. The anion exchange membrane (AEM) produced the largest power density of 610 mW/m^2 and a relatively high C_E = 72%. In comparison, a Nafion membrane produced 514 mW/m^2. Direct comparisons of this two-chamber system are not possible with the 4-cm cube reactor in tests where a membrane is omitted as the conditions are different because the anode chamber is only 2 cm long (14-mL volume compared to 4 cm long, 28-mL volume). Kim *et al.* found no effect of the Nafion on internal resistance as long as the membrane was not placed against the cathode as Nafion is a highly proton conductive material. Recall that in previous tests with the 4-cm cube system that hot-welding the Nafion to the cathode decreased power.

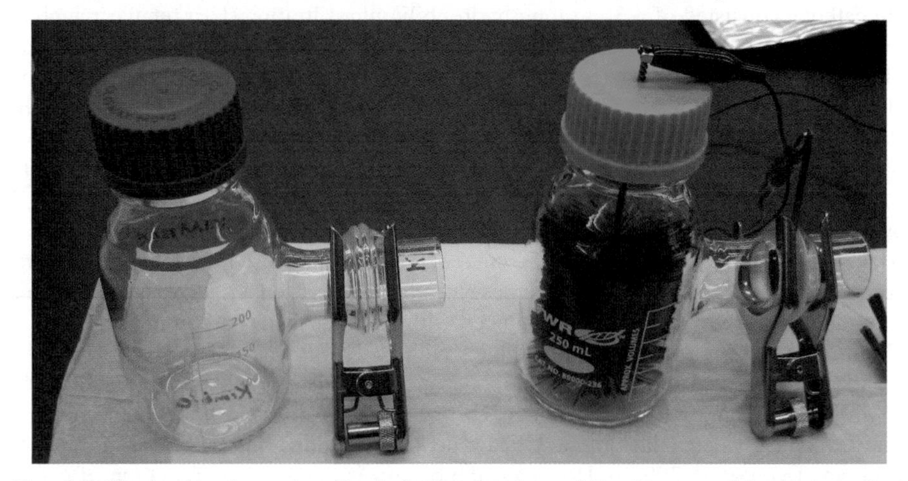

Fig. 6.5 Single-chamber, air-cathode bottle reactors of the type used by Logan *et al.* (2007) with a single side arm shown empty (left) and with a brush anode inside the bottle and carbon paper cathode clamped in the side arm, with the circuit connected with two wires and a resistor.

New bottle reactor MFC. More experimental work is being done with pure cultures, necessitating a system which can be easily autoclaved and kept sterile. A bottle type of single-chamber, air-cathode reactor was developed and tested using both pure and mixed cultures (*Logan et al.* 2007). The reactor consists of a standard laboratory media bottle modified to contain a side arm—essentially one bottle of a two-chambered reactor, with the side arm of a larger diameter than typically used (Fig. 6.5). The side arm is made with a larger diameter to increase oxygen transfer to the cathode, which is held between the side arm and a small additional tube extension. It is shorter to reduce internal resistance and increase proton transfer through the liquid to the cathode. The carbon cloth cathode containing the catalyst on the water facing side (Pt or CoTMPP) is treated with diffusion layers on the air-side to reduce water leakage and evaporation (*Cheng et al.* 2006a). Using graphite brush electrodes (4200 m^2/m^3), the reactor produced a maximum power density with glucose of 1430 mW/m^2 (2.3 W/m^3, C_E = 23%). Power generation was lower using plain carbon electrodes (600 mW/m^2) or random graphite fibers (1100 mW/m^2).

Tests were also conducted using *Shewanella oneidensis* MR-1 and lactate, which produced 770 mW/m^2 with brush electrodes. The side arm and diffusion layers both help to reduce contamination of the culture, and a foam plug can be added to the side arm end to decrease the chance for contamination. However, if the solution is replaced in a glove box extreme care is needed to avoid contamination as the glove box is not sterile and the reactor top cannot be efficiently sterilized. For facultative anaerobes such as *S. oneidensis* which can tolerate and use oxygen, using a laminar flow hood with filter-sterilized air in the laboratory will help reduce the potential for reactor contamination. Before operating the reactor again, however, it must be flushed with filter-sterilized gas to prevent reactor contamination.

Fig. 6.6 Single-chamber air-cathode MFC with a cathode tube concentric with eight graphite anodes in an acrylic tube casing. This reactor was used to demonstrate electricity production with simultaneous wastewater treatment [*Liu et al.* (2004)].

SCMFC. For several years it had been shown that electricity could be generated from wastewater, but the first demonstration of successful wastewater treatment (*i.e.*, COD or BOD) with electricity generation was demonstrated by Liu *et al.* (2004) using a novel MFC reactor design. They constructed a tubular-type of single-chamber, air-cathode MFC that had a cathode inserted in the center of an acrylic tube, with eight graphite rods arranged in a concentric matter around the cathode (Fig. 6.6). The cathode was made by hot pressing a CEM (Nafion) to carbon cloth, and wrapping the cloth around a tube drilled with holes to allow oxygen transfer to the cathode surface. The eight rods were connected to the cathode with a single wire. The reactor removed 80% of the COD and generated 26 mW/m^2 (C_E < 12%). The amount of power was proportional to the hydraulic retention time (HRT), which was examined over a range of 3 to 33 h, and the wastewater strength of the feed (50 to 220 mg/L). The maximum power generated here is similar to that found with the cube reactor (28 mW/m^2) in tests where the cathode and CEM were also bound together, suggesting that power generation in this system was limited for the same reasons observed in the cube reactor system. This reactor design was not tested with other substrates.

Flat Plate MFC. It is well known that hydrogen fuel cells produce substantially more power than MFCs, in part due to the low ohmic resistance as a result of the close spacing of the electrodes. In an effort to produce more power from an MFC, Min and Logan (2004) developed a flat-plate MFC with a CEM (Nafion) sandwiched in between the anode and cathode (Fig. 6.7). The plates (15-cm width by 15-cm length by 2-cm height) were cut to provide a serpentine flow path through the system, providing an approximation of a plug flow type of reactor. The flow path was 0.7 cm wide and 0.4 cm

Fig. 6.7 Flat-plate MFC designed by Min and Logan (2004) that has a CEM clamped between the two electrodes, with the plates drilled to contain a serpentine channel to allow flow of wastewater (WW) and air on opposite sides of the electrodes.

deep, producing a total surface area of 55 cm^2 and a total volume of 22 cm^3, resulting in a specific surface area of 250 m^2/m^3. This reactor produced 56 ± 0 mW/m^2 from domestic wastewater (246 mg-COD/L) and removed 58% of the COD, at a hydraulic retention time (HRT) of 2.0 h (0.22 mL/min flow rate; 164 mg/L log mean COD) with an air flow rate of 2 mL/min (fixed resistance of 470 Ω). Based on power density curves, the maximum power density was 63 mW/m^2 at a current of 1.03 mA (326 Ω) with domestic wastewater. Maximum power densities achieved with other substrates were: 286 mW/m^2, acetate; 242 mW/m^2, starch; 220 mW/m^2, butyrate; 212 mW/m^2, glucose; and 150 mW/m^2, dextran. C_Es for the starch, glucose, and dextrans were lower (14% to 21%) than with acetate (65%).

Power densities produced with these substrates in the flate plate reactor were lower than that achieved with the cube reactor with and a much larger (2-cm) electrode spacing, providing evidence that when the electrodes are too closely spaced, power is not optimized. It is likely that oxygen transfer through the CEM reaches bacteria growing on

Fig. 6.8 Single-chamber MFC using a novel Fe^{3+}-graphite cathode (right) and a more conventional two-chamber system containing a porous ceramic disc separating the two chambers instead of a membrane (left), as developed by Park and Zeikus (2003). (Reprinted with permission from Wiley-Liss Inc., a subsidiary of John Wiley & Sons, Inc.)

the anode, thus reducing the electrogenic growth (*i.e.,* the flow of electrons to the anode) as a result of the more energetically favorable aerobic growth.

Single-chamber reactor with graphite-metal cathode. Park and Zeikus (2003) examined power generation using a single-chamber system with a Fe^{3+}-graphite cathode exposed to air on one side, along with the anode solution on the other side (Fig. 6.8). Several different anodes were tested, with Mn^{4+}-graphite producing the most power compared to plain woven graphite and woven graphite containing a mediator (neutral red). A sewage sludge inoculum produced up to 788 mW/m^2 (34 W/m^3) of anode surface area while an *E. coli* suspension with a soluble mediator (neutral red) produced 91 mW/m^2. Although the cathode was exposed to air, it is not clear whether this was truly a reactor sustained by oxygen or whether iron reduction contributed to power generation. The iron on the cathode was claimed to be chemically regenerated (Fe^{3+} from Fe^{2+}) by oxygen in the air, but it is also possible that the iron functioned as a sacrificial cathode so that the performance of the cathode would have eventually been exhausted as Fe^{2+} dissolved from the cathode into solution. No measurements of iron concentrations in the aqueous chamber or the system longevity were reported.

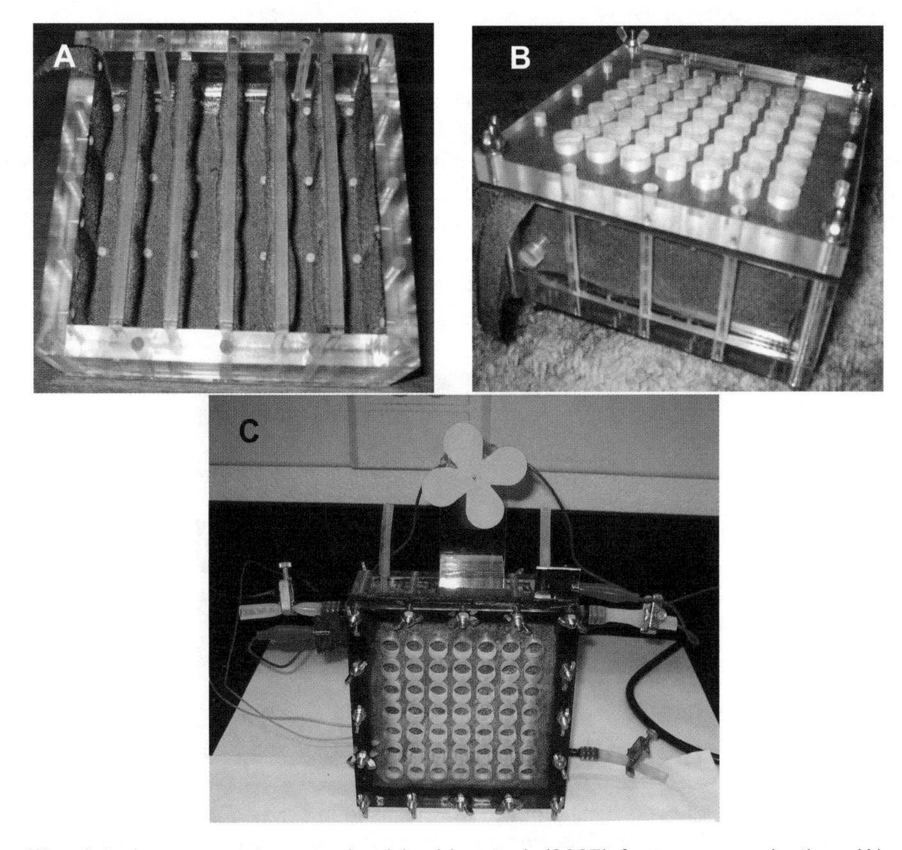

Fig. 6.9 Larger reactor examined by Liu *et al.* (2007) for power production. (A) Anode, showing carbon cloth layered over baffles. (B) Cathode covered with a thick plexiglass cover with holes drilled to allow oxygen to reach the cathode. (C) Reactor powering a fan.

Larger reactors. A relatively large (520 mL) air-cathode MFC (Fig. 6.9) was examined for power production by Liu *et al.* (2007), modeled around the single-chamber air-cathode design used in previous tests by these researchers (*Liu et al.* 2005a, b; *Liu and Logan* 2004). Based on smaller reactor tests and the relative cost of materials, they determined that an anode:cathode ratio based on projected areas of carbon cloth materials and Pt catalysts was optimal if in the range of 4–10. Therefore, they designed and tested a reactor with a ratio of 4.7 which was achieved by winding a carbon cloth over a series of baffles. This reactor produced a maximum of 22 W/m^3 (695 mW/m^2 normalized to cathode projected surface area) when operated in a continuous mode at a hydraulic retention time (HRT) of 16.9 hours. Over 90% of the substrate in the feed (acetate) was removed. The power production is similar to that achieved with the smaller reactor, although it was not as large as predicted based on the chosen electrode surface area ratios. The reactor is currently being operated in batch mode, achieving a constant power output of 10 mW. The reactor is fed every few days with a sodium acetate solution, but sometimes it runs out of fuel and is not immediately re-fueled. Even under these conditions, it has run for over one year, and it can be viewed continuously at www.engr.psu.edu/mfccam.

A second large-format reactor built by Penn State researchers was built around the concept of using graphite granules to provide a conductive surface area well in excess of the surface area of the cathode (Fig. 6.10). This reactor consisted of two MFCs connected in series, with the cathodes facing the outside and the granular reactors on the inside. Each chamber contains 0.5 kg of graphite granules 2 to 6 mm in diameter, a cathode 182 cm^2 in projected surface area, and 0.5 L of a buffered nutrient medium containing sodium acetate. The maximum power produced by the system is 2000 mW/m^2 (cathode surface area). Each anode chamber contains a vent to allow CO_2 gas produced from acetate degradation to escape. This system, like the one described above, has run for over a year and can be viewed online in real time at www.engr.psu.edu/mfccam.

Fig. 6.10 A larger MFC consisting of two MFCs linked in series. Each anode chamber is filled with 0.5 L of solution. The system produces up to 2000 mW/m^2 of power from acetate, and it runs a small fan.

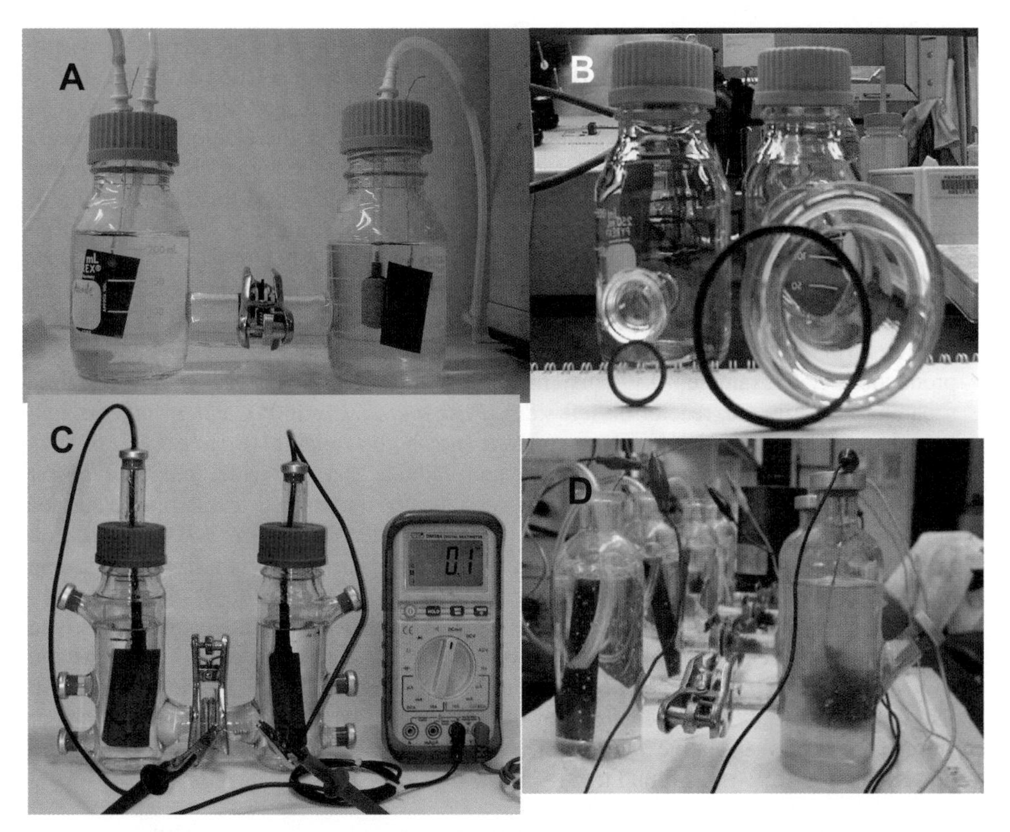

Fig. 6.11 Different H-type reactor MFCs. (A) Both chambers can be sparged with gas—for example, the anode with nitrogen to maintain anaerobic conditions, and the cathode with air. [Reprinted from *Logan et al.* (2006), with permission from the American Chemical Society.] (B) The CEM size can be increased, resulting in greater power generation. [For example, see *Oh and Logan* (2006).] (C) Reactors set up to maintain anaerobic conditions in both chambers, with several sampling ports all sealed with thick rubber stoppers. [Modified from *Lovley* (2006), reprinted with permission, Nature Publishing Group.] (D) Simpler design for a crimp-top bottle with a single sampling side port on the anode chamber.

6.3 Aqueous cathodes using dissolved oxygen

The simplest MFC design consists of two chambers separated by a cation exchange membrane (CEM). Many researchers use such systems to investigate power production (*Bergel et al.* 2005; *Bond et al.* 2002; *Min et al.* 2005a; *Oh and Logan* 2005; *Oh et al.* 2004), but there are few photographs of these systems in the literature. The distinguishing features of these systems are two chambers, the CEM, and a cathode suspended in water that is sparged with air (Fig. 6.11). The concentration of dissolved oxygen can affect performance (*Oh et al.* 2004), with power decreasing as DO is lowered or increasing if pure oxygen is used. The anode chamber can also be sparged with gas (for example, with nitrogen to remove oxygen), although some bacteria may be adversely affected by the shear created by the gas bubbling (Fig. 6.11A). The chambers can be mixed with a stir bar to maintain homogeneous conditions in the reactor.

The size and proton transfer efficiency of the CEM connecting the chambers usually limits power production in these H-type systems. Because the CEM is permeable to oxygen and to substrate and other chemicals in the solution, providing a larger CEM sometimes helps increase power generation but it may lower the Coulombic efficiency due to loss of substrate or increased transfer of oxygen into the anode chamber (*Oh and Logan* 2006) The use of the system shown in Fig. 6.11A produced the same amount of power (35–45 mW/m^2) independent of the inoculum or substrate in several studies (*Kim et al.* 2007c; *Logan et al.* 2005; *Min et al.* 2005a; *Min et al.* 2005b; *Oh and Logan* 2005; *Oh and Logan* 2006; *Oh et al.* 2004). The high internal resistance of this system (~1300 Ω) is caused by the large electrode spacing and the ineffective reaction of dissolved oxygen at the catalyst. Other two-chamber systems have similarly observed low power production—for example, 0.17 mW/m^2 in a two-chamber system with a ceramic separator (see Fig. 6.8, reactor on left side) (*Park and Zeikus* 2003), and 14 mW/m^2 in a two-chamber system with a Nafion membrane (no photo available) (*Bond et al.* 2002).

By choosing different CEM, anode, and cathode sizes, a wider range of power densities is possible. For example, the power density normalized to the anode surface area increased with the CEM size in the following order: 45 mW/m^2 (A_m=3.5 cm^2), 68 mW/m^2 (A_m = 6.2 cm^2), and 190 mW/m^2 (A_m = 30.6 cm^2) (all with fixed anode and cathode surface areas of A_{An}= A_{Cat} = 22.5 cm^2) (*Oh and Logan* 2006). As shown in Fig. 4.1, as the relative sizes of the electrodes changes, the power output changes. Thus, you can make the power production appear to be large by increasing the size of the CEM and cathode relative to that of the anode if you normalize power by the anode surface area. Oh and Logan (2006) showed that power was quite predictable in their two-chamber system based on the relative sizes of the CEM and the electrodes, where the maximum power output P (mW) was found to be

$$P = \frac{A_m}{10{,}000} \times \frac{126.6 \times \left(A_{Cat} / A_m\right)^{0.439}}{1 + \left[0.155 \times \left(A_m / A_{An}\right)\right]^{2.45}} \tag{6-1}$$

where areas have units of cm^2. The fit of this equation to experimental data was shown in Fig. 4.1. Note that this is an empirical result that only applies to the system being used, and it cannot be applied to other systems. However, it is a good example of how the response of the system is predictable based on changes in architecture due to surface areas and CEM sizes.

CEMs do not need to be used in the two-chamber reactor. For example, Min *et al.* (2005a) used a salt bridge consisting of a tube filled with agar and salt and capped with porous caps (Fig. 6.12A). They produced a very small amount of power, however, achieving only 2.2 mW/m^2 in a reactor inoculated with the iron-reducing bacterium, *G. metallireducens*. The low power output was a result of the very high internal resistance (19,920 Ω) which they measured using electrochemical impedance spectroscopy (EIS).

Tubes are not needed to connect the anode and cathode chambers. With the right design, the two chambers can be pressed against one another but kept separated with a CEM or UF membrane. We shall see several examples of these types of reactors tested using chemical catholytes (but not oxygen) in the examples below. One novel design using dissolved oxygen was a U-shaped design developed by Millikan and May (2007) (Fig. 6.12B). Two tubes were directly joined together with a CEM separating the chambers, with the electrodes suspended in solution on either side of the CEM. This design avoids having the electrodes right on top of the CEM (see Fig. 6.7) and allows the top of the anode tube to be sealed with a thick rubber stopper for working with a pure culture. The reactor was tested for power production using a gram-positive culture of *Desulfitobacterium hafniense* strain DCB2, and produced up to 400 mW/m^2, but a mediator (AQDS) was needed for this strain.

The fact that power generation is so sensitive to the sizes of the electrodes, CEM, electrode spacing, and other factors makes it essentially impossible to know if one bacterium can produce more power than another, or if power is affected by substrate of medium composition, unless all variables are tested in the same system—and power is not limited by internal resistance due for example to the CEM. In addition, the use of different catholytes (or poising the potential) will similarly affect power. Therefore, we must be critical of claims that bacteria produce more or less power than other strains unless conditions are exactly matched to rule out other factors.

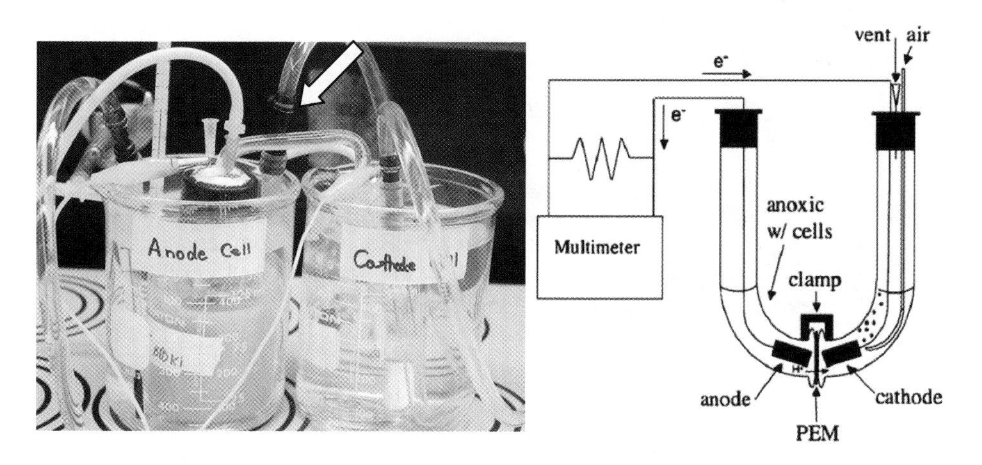

Fig. 6.12 Variations on two-chamber reactors. (A) A salt bridge-type MFC used by Min *et al.* (2005a) between the chambers (arrow), but the system had high internal resistance resulting in low power densities. [Reprinted from *Logan et al.* (2006), with permission of the American Chemical Society.] (B) U-shaped MFC with the two chambers joined directly with a clamp [From *Milliken and May* (2007), reprinted with kind permission from Springer Science and Business Media .]

6.4 Two-chamber reactors with soluble catholytes or poised potentials

Ferricyanide catholytes. In the quest to provide the most power from an MFC, the leader in the race remains the two-chamber MFC developed by Rabaey *et al.* (2004) which produced up to 4310 mW/m^2 (based on projected surface area, 253 W/m^3) with a glucose substrate, slightly greater in power than that achieved with the same system in an earlier study by this group (3600 mW/m^2, 216 W/m^3) (*Rabaey et al.* 2003). This remarkable accomplishment was achieved using an acclimated consortium of bacteria that produced mediators, in a reactor that had the electrodes fitted into the chambers near the CEM (Fig. 6.13). The use of ferricyanide accomplished efficient electron transfer at the cathode, and the internal resistance of this system must have been low to achieve this high power

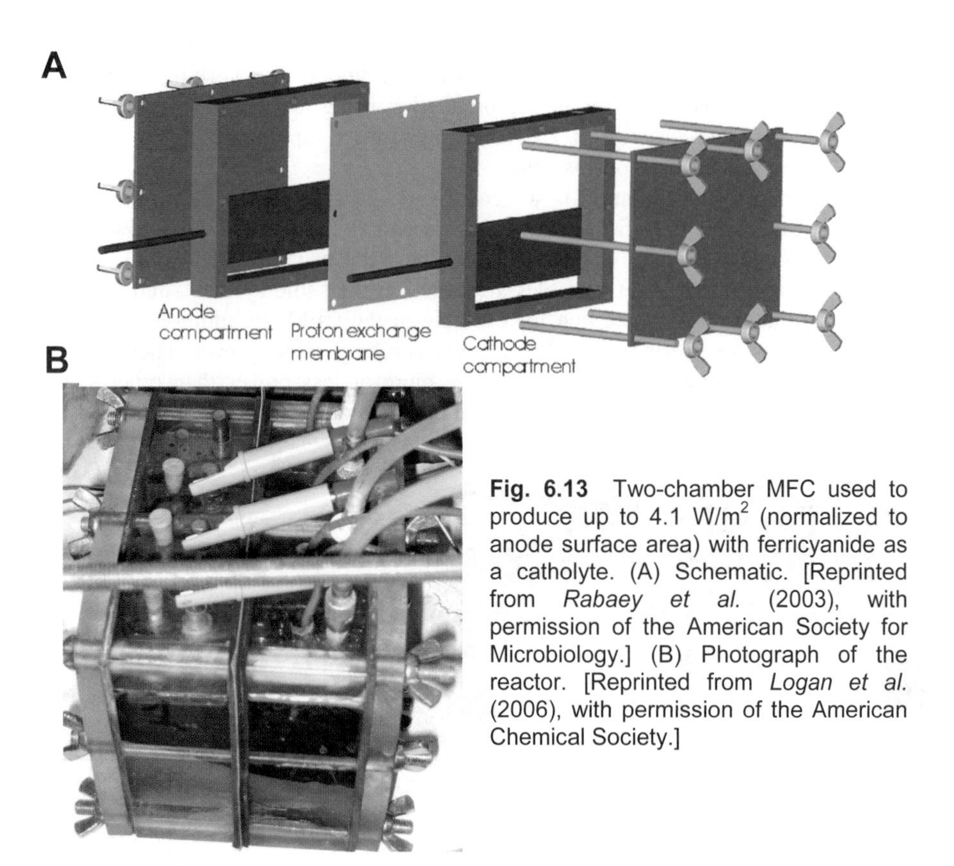

Fig. 6.13 Two-chamber MFC used to produce up to 4.1 W/m^2 (normalized to anode surface area) with ferricyanide as a catholyte. (A) Schematic. [Reprinted from *Rabaey et al.* (2003), with permission of the American Society for Microbiology.] (B) Photograph of the reactor. [Reprinted from *Logan et al.* (2006), with permission of the American Chemical Society.]

density. Graphite rods set into graphite plate electrodes (roughened to increase surface area) provided good electrical contact, and the highly conductive graphite material was highly conductive for facilitating electron transfer from bacteria and mediators.

Ferricyanide alone does not guarantee a high power output, but it does typically increase power output compared to systems using dissolved oxygen at the cathode. The replacement of the air-sparged water with a ferricyanide solution increased the maximum power by 50–80% in a standard two-bottle reactor (Fig. 6.11A), but power was still quite low (43 mW/m^2 anode surface area) (*Oh et al.* 2004). Zhang *et al.* (2006a) also used ferricyanide in a two-bottle type of MFC, producing 50 mW/m^2 with a glucose substrate, and 70 mW/m^2 with acetate. Thus the use of ferricyanide alone is not sufficient to achieve high power densities if the electrode spacing is large and the CEM size is small.

Micro reactor with ferricyanide and an external feed. An unusually small reactor was developed by Ringeisen *et al.* (2006), consisting of two chambers each 1.2 cm^3 in total volume (Fig. 6.14). The cathode contained ferricyanide as a catholyte, with the solution exposed to air on one side of the reactor. A suspension of *Shewanella oneidensis* DSP10 was grown externally on lactate in a 100-mL flask under poorly defined micro-aerobic conditions. The flask was not mixed but the suspension was exposed to oxygen in air, with the bacteria in suspension consuming oxygen and presumably creating anoxic conditions in the culture. The cell suspension was pumped through the small MFC,

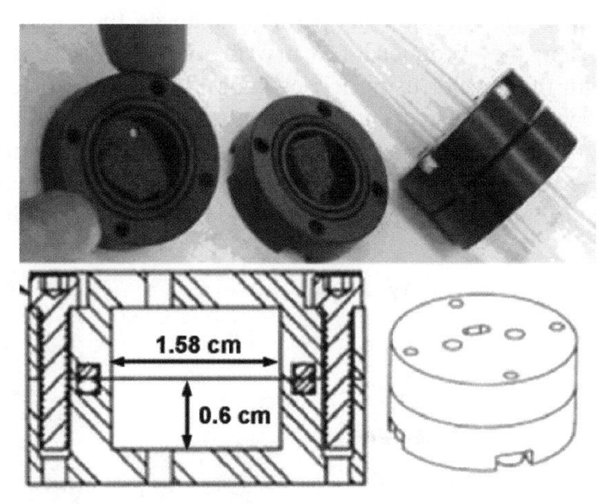

Fig. 6.14 Mini-MFC (1.2 cm^3 each for the anode and cathode) developed by Ringeisen *et al.* (2006) that was used to measure continuous power production by pumping a suspension of *Shewanella oneidensis* DSP10 grown in a 100-mL flask and pumped through the reactor. [Reprinted from *Ringeisen et al.* (2006) with permission of the American Chemical Society.]

creating 10 mW/m^2 (graphite felt anode) or 24 mW/m^2 (RVC anode) depending on the anode material. These surface areas are the actual areas determined by the authors, which for the case of the graphite felt were 60 times the projected surface area. Based on the projected anode surface area or reactor volume, this system produced 3 W/m^2 and 500 W/m^3. However, these values do not consider the volume of the external reactor, which would substantially reduce this volumetric power density. The mechanism of power production in this system is not fully understood, but it was thought that direct contact of the cells with the anode material provided sufficient time for electron transfer to the anode. It was not thought that mediators were present in the system, and in fact the addition of an exogenous mediator (AQDS) increased power by 30–100%.

Permanganate catholyte. A simple H-type of reactor made of two plastic bottles and

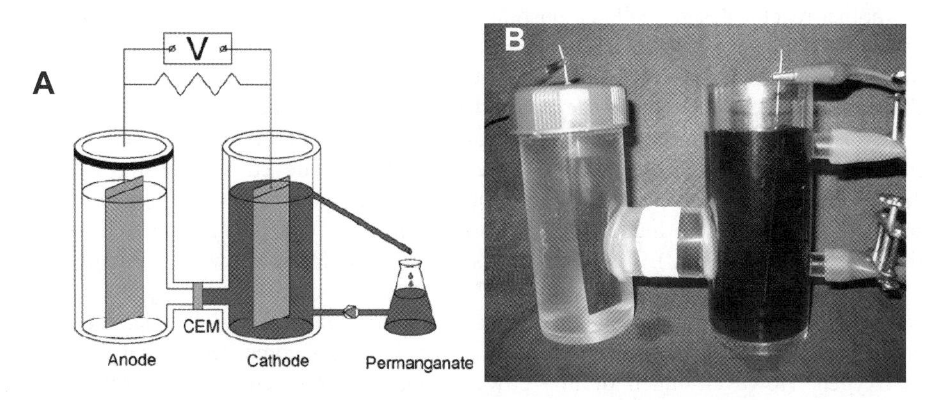

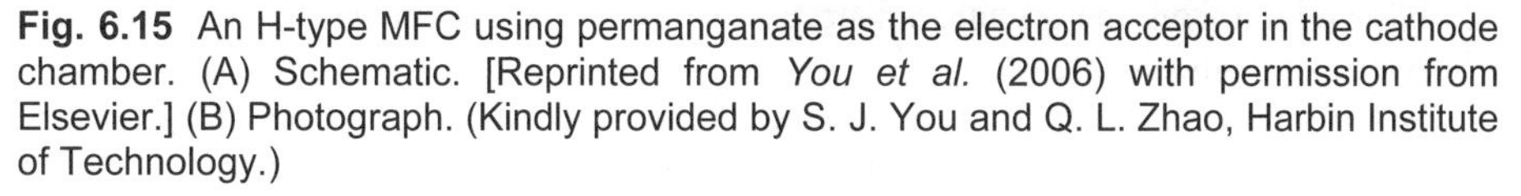

Fig. 6.15 An H-type MFC using permanganate as the electron acceptor in the cathode chamber. (A) Schematic. [Reprinted from *You et al.* (2006) with permission from Elsevier.] (B) Photograph. (Kindly provided by S. J. You and Q. L. Zhao, Harbin Institute of Technology.)

carbon paper (anode) and carbon cloth (cathode) was examined for power production by You *et al.* (2006) with permanganate as the electron acceptor (Fig. 6.15). With this system they produced 116 mW/m^2, which was substantially higher than that achieved with this reactor using ferricyanide (26 mW/m^2) or dissolved oxygen (10 mW/m^2). The large power density with permanganate was a result of the higher cathode potential as permanganate has a standard potential of 1.70 V compared to 0.361 V for ferricyanide (see Table 3.1). The *OCP* produced by this system was 1.38 V, demonstrating the contribution of the chemical energy in permanganate on power production.

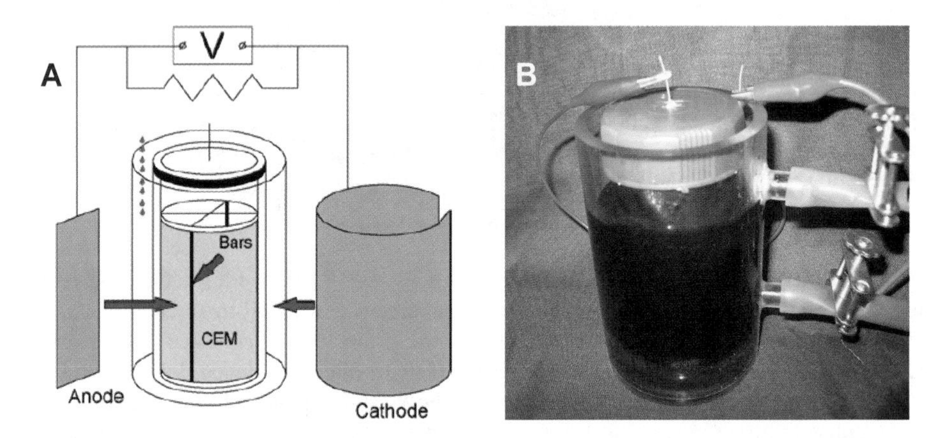

Fig. 6.16 Two-chamber "bushing" reactor developed by You *et al.* (2006) that used permanganate as the catholyte. (A) Schematic. [Reprinted from *You et al.* (2006), with permission from Elsevier.] (B) Photograph. (Kindly provided by S. J. You and Q. L. Zhao, Harbin Institute of Technology.)

Another reactor design, called a bushing MFC, was tested by You et al. (2006) again with a permanganate catholyte. This reactor consisted of a plastic bottle (3.5 cm in diameter, 10 cm high) containing four bars that held a CEM hot-pressed onto the bars, with a cathode placed in a concentric a cylindrical shape inside the bottle (Fig. 6.16). The anode was plain carbon paper. The reactor produced 3990 mW/m^2 of power with permanganate, which is substantially larger than that achieved with a ferricyanide catholyte (1230 mW/m^2). The contribution of the chemical energy to this power density is again evidenced by the higher *OCP* with permanganate (1.532 V) compared to that with ferricyanide (0.788 V). The internal resistance was also lower with permanganate (51 Ω) compared to that with ferricyanide (73 Ω). The use of a dissolved oxygen cathode was not reported in this study. Thus it is clear that permanganate, ferricyanide, and other chemical catholytes can be used to increase power in MFCs compared to that achieved with oxygen. However, much of this power is derived from chemical energy in the catholyte and not from the organic substrate, and unlike oxygen the use of these chemical catholytes is not sustainable.

Poised potential reactors. In a reactor with oxygen as the electron acceptor at the cathode, the working potential is typically *ca.* 0.25 V, with the anode at *ca.* −0.2 V, producing 0.45 V (see Fig. 3.1). In a poised potential experiment, the anode potential can be set at any value using a potentiostat. Once that is done, the bacteria "see" a terminal electron acceptor (the anode) at a specific potential, allowing transfer of those electrons to a surface at a known potential. This has several advantages, one of which is that the potential of the anode is now precisely known. The other one is that the system can be run under a completely anaerobic condition as oxygen is not needed at the cathode. With the anode potential set, the circuit voltage will drive hydrogen gas production at the cathode (see Chapter 8 for a more detailed discussion on this phenomenon). Thus, the "electron acceptor" in this case is H^+, making it unnecessary to use oxygen or chemicals such as ferricyanide.

When a poised potential is used, energy is put into the system. For example, if the working potential for the anode was −0.3 V but the potential was poised at 0.2 V, then the input power would be $P = 0.5\ I$, where I is the current achieved in the experiment. Thus, power is input into this system directly in a non-sustainable manner, as opposed to indirectly using the chemical energy of catholytes such as ferricyanide. Another potential disadvantage of this set up is that H_2 gas produced in the cathode chamber. If this gas diffuses back into the anode chamber, it can become another substrate for the bacteria which can artificially raise the Coulombic efficiency or power production.

Bond *et al.* (2002) used poised potentials to examine current generation by *Desulfuromonas acetoxidans* and *Geobacter metallireducens* in a two-chamber system, finding 82% and 84% of electrons were accounted for as current using acetate and benzoate, respectively. These results seem reasonable given the absence of other terminal electron acceptors and the use of pure cultures. Chaudhuri and Lovley (2003) found a Coulombic efficiency for current production from glucose by *Rhodobacter ferrireducens* was 83%. However, the finding of a Coulombic efficiency of 96.8% by *Geobacter sulfurreducens* seems unreasonably large, suggesting that either substrate had been accumulated in the bacteria (*Freguia et al.* 2007) and used during these tests or that H_2 gas was leaking from the cathode into the anode chamber. Otherwise, the biomass yields by this microbe would be unusually low.

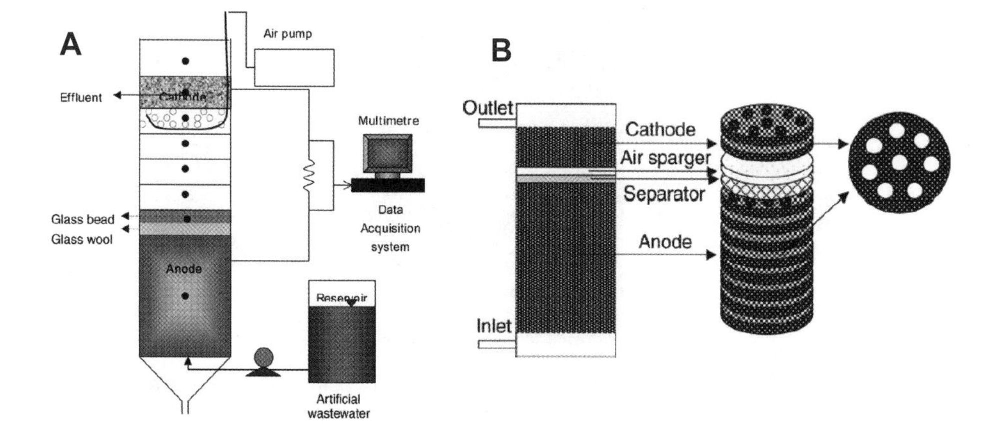

Fig. 6.17 Continuous-flow tubular reactors developed by (A) Jang *et al.* (2004), and (B) Moon *et al.* (2005) for electricity generation. The flow moves from the anode chamber into the cathode chamber where it is sparged with air. (Reprinted with permission from Elsevier.)

6.5 Tubular packed bed reactors

Several researchers have now investigated the use of reactors in a tubular shape that contains a conductive packing and are specifically designed to run in a continuous flow mode. Liu *et al.* (2004) used a tubular MFC that contained eight graphite rods and a central cathode tube, as described above (Fig. 6.6). Some of these reactors use oxygen at the cathode, while others rely upon the use of chemical catholytes such as ferricyanide for power production.

Jang *et al.* (2004) used a novel tubular reactor approach based on the flow moving through an anode chamber and then directly into the cathode chamber in the same column (Fig. 6.17A). The cathode chamber is sparged with air to provide oxygen at the cathode. In their first reactor design, the two chambers were separated by glass wool and glass beads (Fig. 6.17A). The reactor produced only 1.3 mW/m^2 with a glucose-glutamic acid solution, and an anode made of a roll of carbon felt and a cathode of the same material placed in disk-form above the air sparger. In a later design a perforated polyacrylic plate was used as a separator, and the cathode was Pt-coated graphite felt cathode was perforated (Fig. 6.17B) (*Moon et al.* 2005). This newer design produced up to 560 mW/m^2, with a volumetric power density of 102 W/m^3 (*Moon et al.* 2006). Power generation was reported to be stable for over two years. One disadvantage of this reactor system is that if the organic matter is not removed sufficiently in the anode chamber it will flow into the cathode chamber and create an oxygen demand or wastewater will leave the reactor untreated. This second-stage oxygenation step may also be an advantage,

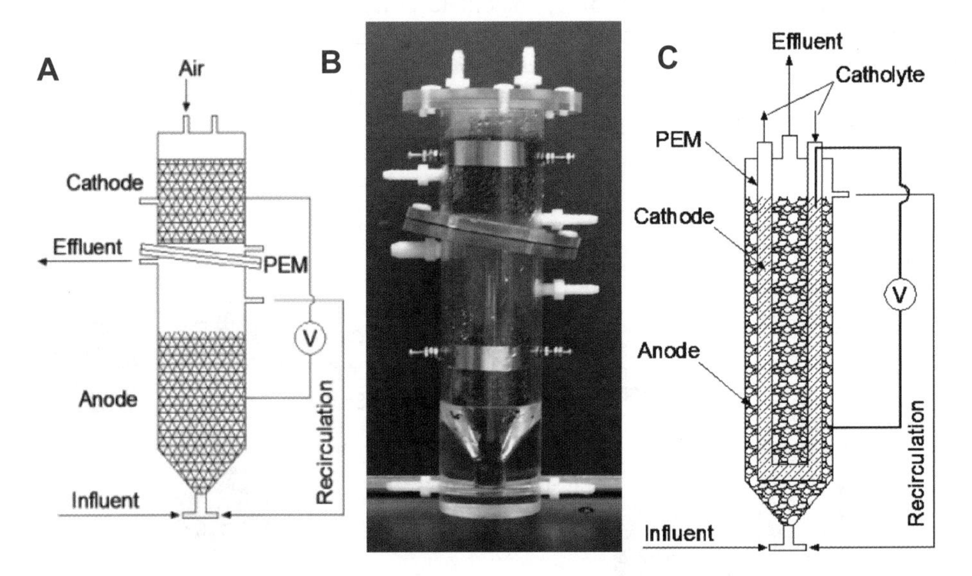

Fig. 6.18 Two different tubular upflow reactor designs using ferricyanide cathodes. (A) Schematic and (B) photograph of an upflow system with the reactor packed with reticulated vitreous carbon (RVC), with the CEM separating the two chambers (*He et al.* 2005). (B) Schematic of a reactor packed with granular activated carbon (GAC) with a tube made of a CEM and packed with GAC, that contains a ferricyanide solution (*He et al.* 2006). (Reprinted with permission of the American Chemical Society.)

however, as additional aerobic treatment can occur reducing the organic content of the effluent. However, substantial energy input is needed for the aeration stage compared to a reactor with an air cathode.

Two different tubular flow reactor designs were developed by the same group (*He et al.* 2005; *He et al.* 2006) based on using a bed of porous and electrically conductive material and ferricyanide solutions. The first reactor developed by He *et al.* (2005) contained an anode packed with reticulated vitreous carbon (RVC) (51 m^2/m^3), with a total volume of 190 mL (Fig. 6.18A,B). The flow was directed towards the CEM (CMI-7000, Membranes International, Inc., Glen Rock, NJ), with the cathode chamber above the CEM also containing RVC (114 m^2/m^3) and a ferricyanide solution. The reactor produced up to 170 mW/m^2 when fed a sucrose solution, with C_E's ranging from 0.7% to 8.1%. The reactor had an internal resistance of 84 Ω, which limited power production and likely contributed to the low C_E. The next upflow reactor design by He *et al.* (2006) contained a CEM formed into a tube that was placed into a bed of granular activated carbon (GAC) (Fig 6.18C). This reactor produced 29 W/m^3 with C_E's ranging from 11% to 51%. This reactor had a lower internal resistance of 17.13 Ω, which the authors separated into an electrolyte resistance of 8.62 Ω, a charge transfer resistance of 7.05 Ω, and a diffusion resistance of 1.46 Ω using a simple equivalent circuit analysis. The soluble COD removal efficiencies exceeded 90% with effluent volatile acid concentrations of ~ 40 mg/L. The authors suggested that to be competitive with anaerobic digestors producing methane gas for electricity generation, that designs were needed that achieved 160 W/m^3 or more of power.

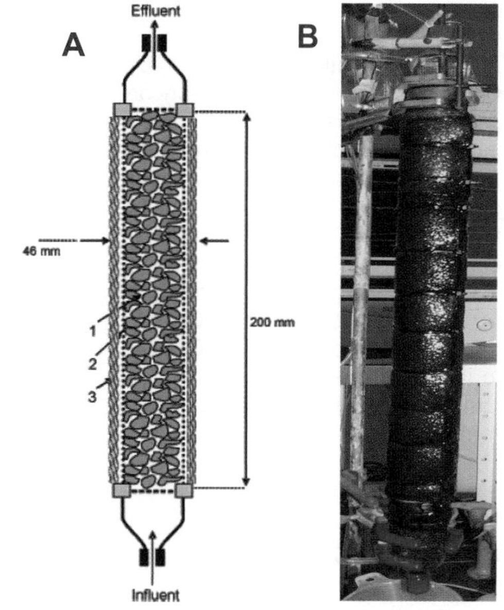

Fig. 6.19 Tubular reactor design where the tube is made of a CEM and contains conductive graphite granules packed into the center of the tube, with the ferricyanide catholyte fluid overflowing the outside of the column which is covered with a thick woven graphite mat. (A) Schematic. [Reprinted from *Rabaey et al.* (2005b), with permission of the American Chemical Society.] (B) Photograph of the reactor. (Provided by K. Rabaey, Ghent University.)

A different tubular design was developed by Rabaey *et al.* (2005b) based on using conductive graphite granules that were packed into a CEM (CMI-7000) soldered to form a tube (Fig. 6.19). The influent flowed up through the packed bed and out the top of the reactor, while the catholyte (ferricyanide) flowed down and over the outside of the reactor covered with a thick woven graphite mat (Alfa Aesar, Belgium) pressed onto the CEM. The packing porosity was 0.53, with an estimated specific surface area of the graphite granules that were 1.5 to 5 mm in diameter of 820 to 2700 m^2/m^3. This reactor produced 48 W/m^3 with aceate and 38 W/m^3 with glucose (based on total reactor volume). When switched to domestic wastewater, the maximum power was 25 W/m^3 with only 20% of the organic matter removed converted to current. Sulfate removal and sulfide production was observed, indicating COD removal via an alternate electron acceptor.

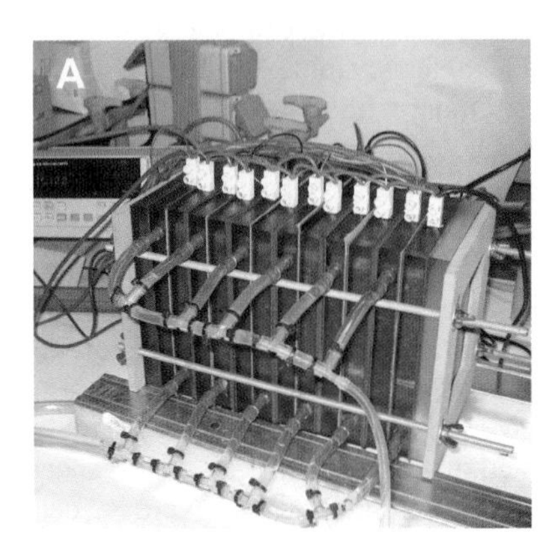

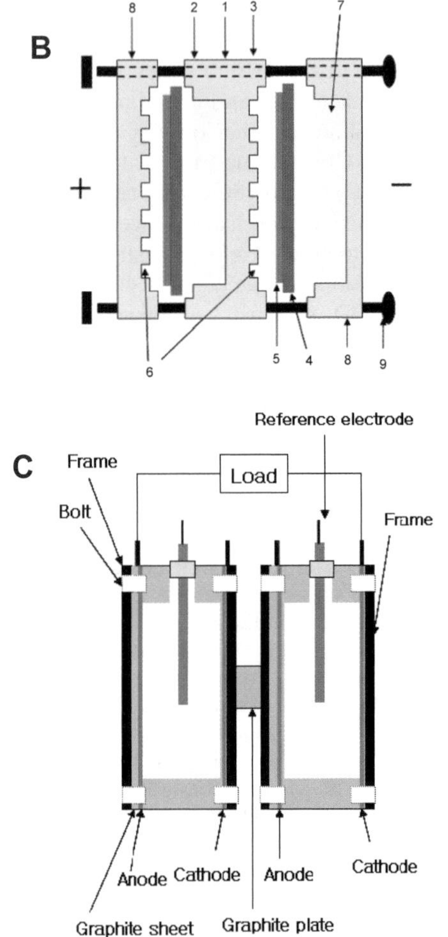

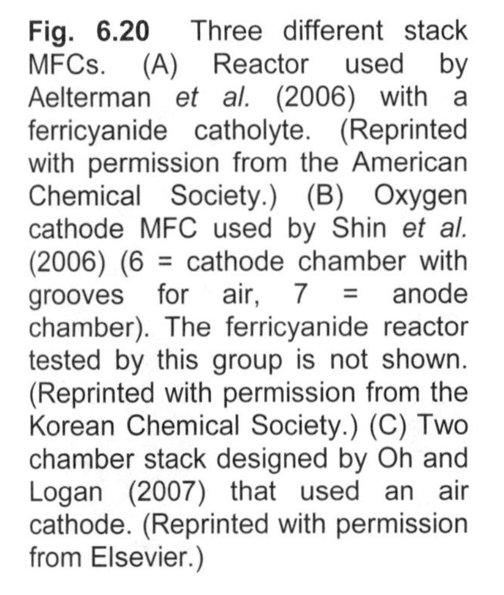

Fig. 6.20 Three different stack MFCs. (A) Reactor used by Aelterman *et al.* (2006) with a ferricyanide catholyte. (Reprinted with permission from the American Chemical Society.) (B) Oxygen cathode MFC used by Shin *et al.* (2006) (6 = cathode chamber with grooves for air, 7 = anode chamber). The ferricyanide reactor tested by this group is not shown. (Reprinted with permission from the Korean Chemical Society.) (C) Two chamber stack designed by Oh and Logan (2007) that used an air cathode. (Reprinted with permission from Elsevier.)

6.6 Stacked MFCs

A single MFC produces a low voltage, but by stacking MFCs together in series it is possible to increase the voltage. There can be losses when individual cells are joined in series so that the final voltage may not be equal to the sum of the individual cell voltages. Stacked MFCs have only been investigated in a few studies, and voltage reversal remains a large obstacle for successful increases in the voltage. Aelterman *et al.* (2006) examined power production in a 6-cell stack with acetate as the substrate and ferricyanide as the catholyte (Fig. 6.20A). Reactors were connected using copper wires in series (or in parallel), and thus were not constructed with bipolar plates as done for hydrogen fuel cells. The MFCs were separated by rubber sheets, with the anode and cathode chambers (each 156-mL total volume, 60- mL liquid volume) containing graphite rods set into beds of graphite granules. The reactor produced 59 W/m^3 in stack (parallel) mode and 51 W/m^3 in series mode based on total volume (6 × 312 mL = 1.9 L). The stack produced 118 W/m^3 when normalized to the total anode volume, and 308 W/m^3 normalized by only the liquid anode. Internal resistances of the cells decreased over time, from 6.5 to 3.9 Ω. The Coulombic efficiency was only 12% when cells were operated as a stack (in series), but increased to 78% when reactors were operated in parallel. In monitoring the individual cells the authors found unequal voltages for the cells, with voltage reversals occurring resulting in some cells having negative voltages. The reasons for voltage reversal were not further investigated.

A bipolar-plate type of stacked MFC containing five cells was investigated by Shin *et al.* (2006) with a chemical mediator (thionin) in tests with *Proteus vulgaris* growing on glucose (Fig. 6.20B). While power densities were not reported on a volumetric basis, the reactors produced 1300 mW/m^2 with ferricyanide as the catholyte and 230 mW/m^2 with pure oxygen gas. They used a 1-F supercapacitor with a 5.5-V limit to store energy produced by the stack so that they could light a small bulb based on an output voltage of 2.3 V when using ferricyanide.

The reasons for voltage reversal in stacks were further investigated by Oh and Logan (2007) using a two-cell stack with acetate as the substrate and oxygen in air as the electron acceptor (Fig. 6.20C). Initially the two cells produced the same power, but after several feeding cycles one cell produced a lower voltage at the beginning of a feeding cycle, and then before the cycle was complete the cell voltage reversed, resulting in the stack voltage decreasing from 0.38 V to 0.08 V. It was shown that substrate depletion could drive one cell into voltage reversal. When the cell undergoing voltage reversal was fed substrate in the next cycle, but the first cell was not, the first cell underwent voltage reversal. Subsequently, the first cell suffered voltage reversal indicating the ability of the bacteria in that cell to produce a higher voltage was damaged.

These findings indicate that linking cells in series to form stacks may be a difficult way to increase voltage as variations of output in individual cells could drive the stack power output to rapidly falter. While chemical fuel cells can be matched closely in power output to avoid voltage reversal, fluctuations in biological systems are more common and thus power generation could be adversely affected. It might be possible to eliminate charge reversal in a cell in a stack using a diode, but more work is needed in this area to demonstrate the feasibility and stability of stack systems. In addition, there does not seem to be good evidence that bipolar plates are needed at the current densities produced by MFCs. The power produced by two cells connected with a plate or with a wire appear similar.

6.7 Metal catholytes

Using electron acceptors other than oxygen can produce power, but chemicals need to be regenerated. Two different approaches have been taken to try to use metals as the electron acceptor in an MFC. One approach used by Shantaram *et al.* (2005) and Rhoads *et al.* (2005) was based on using a solid-state cathode made of manganese oxide. MnO_2 is

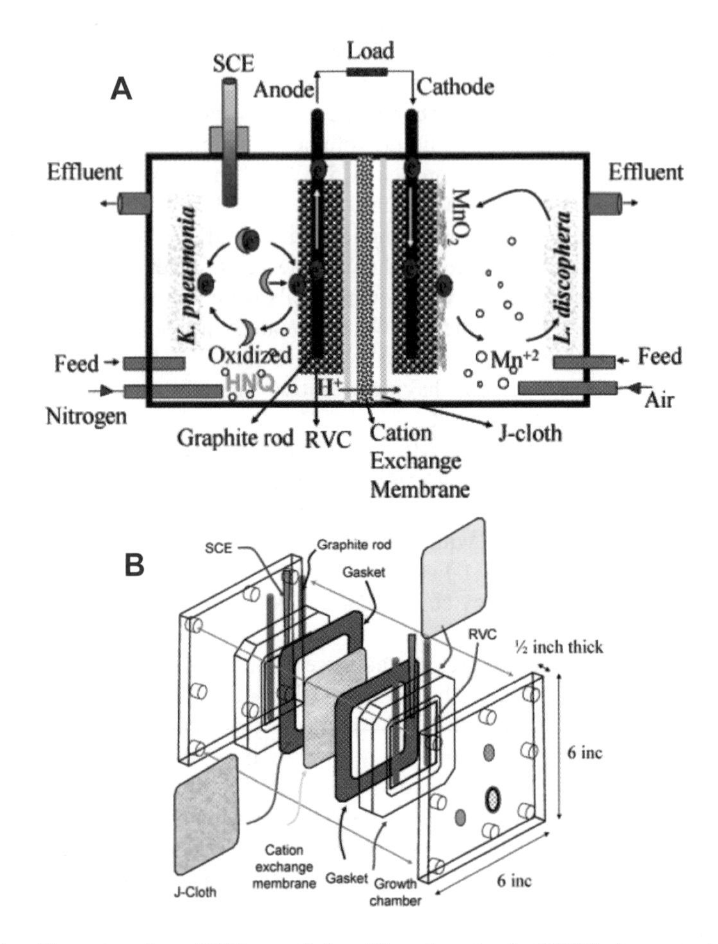

Fig. 6.21 Two-chamber MFC used by Rhoads *et al.* (2005) to examine power production using biomineralized manganese in the cathode chamber by a pure culture of *Leptothrix discophora*. (A) Overall process occurring in the MFC. (B) Schematic of the reactor. [Reprinted from *Rhoads et al.* (2005) with permission of the American Chemical Society.]

reduced to MnOOH via electrons from the cathode, releasing Mn^{2+} into solution. Under proper conditions the Mn^{2+} is oxidized by manganese-oxidized bacteria back to MnO_2 using oxygen as the electron acceptor. It was shown by Rhodes *et al.* that current generated by oxidation of glucose by *Klebsiella pneumoniae* in the presence of an exogenous mediator (2-hydroxy-1,4-haphthoquinone; HNQ) was sustained using the solid-state cathode inoculated with a manganese oxidizing bacterium *Leptothrix discophora* in a two-chamber MFC using RVC electrodes (Fig. 6.21). A peak power output of 127 mW/m^2 was obtained, but it was not demonstrated how long power could be sustained by this approach. Additional work is needed to prove that the manganese was not just serving as a sacrificial electrode material, and that power could be sustained through biocatalysis at the cathode. Also, oxygen must be dissolved in water in the

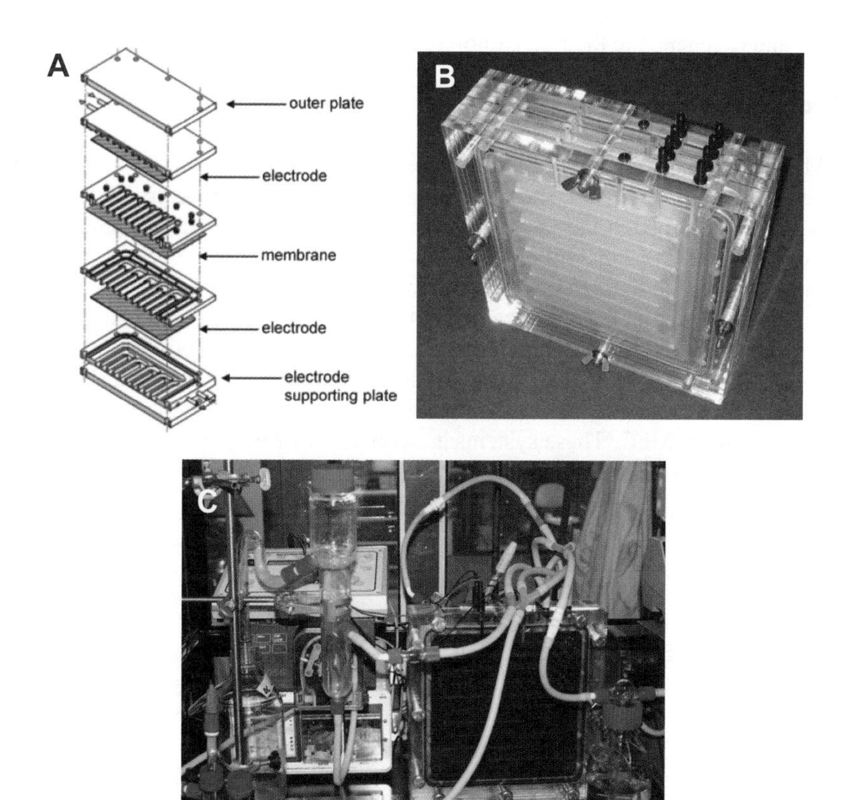

Fig. 6.22 Flat-plate reactor used with a soluble iron catholyte and a bipolar membrane developed by ter Heijne *et al.* (2006). (A) Schematic of reactor. (Reprinted with permission of the American Chemical Society.) (B) Photograph of a clean reactor. (C) Reactor in operation with recycle lines for anode and cathode solutions. (Kindly provided by A. ter Heijne, WETSUS.)

cathode chamber for this system to work, and aeration is an energy demanding process if done in an engineered system.

A second approach to using a metal catholyte, used by ter Heijne *et al.* (2006), was based on using iron dissolved in solution. The pH of the iron catholytes solution was kept low (pH = 2.5) to keep the iron dissolved. A bipolar membrane was used to separate the cathode chamber from the anode chamber which operated at a pH of 6 or 7. The reactor construction was similar to that of Min and Logan (2004) and consisted of a flat-plate arrangement with the bipolar membrane sandwiched in between the two electrodes made of graphite felt (Fig. 6.22). The catholyte solution containing iron chloride or iron sulfate was continuously recirculated through the cathode chamber, and had a dissolved oxygen concentration of 20% air saturation. The anolyte containing acetate was also recirculated through the anode chamber to avoid diffusion limitations to the bacteria. The reactor produced 860 mW/m^2 (42 W/m^3), with a C_E reaching 80–95% and an energy efficiency based on acetate consumption of 18–29%. However, the energy needed to recirculate the solutions and to regenerate the iron catholyte solution was not included in this analysis.

6.8 Biohydrogen MFCs

Most MFC reactors that have been considered so far consist of studies where power is generated by exoelectrogenic bacteria, although in a few cases reactors described above were included even though they were tested with mediators in order to highlight a novel reactor design. These reactors all operate on the principle of direct electron transfer to an electrode by either the bacteria or a mediator. A different approach has been used by Schröder and co-workers based on using bacterial fermentation to generate H_2 gas, with the H_2 reacting on the anode containing a chemical catalyst to produce electrons and protons. These types of H_2-based systems are really hydrogen fuel cells as hydrogen reacts at the anode, but because bacteria are present in the system, they can also be considered a type of MFC. These systems are collectively referred to here as biohydrogen MFCs (HMFCs) in order to distinguish the H_2 gas reaction occurring at the anode from

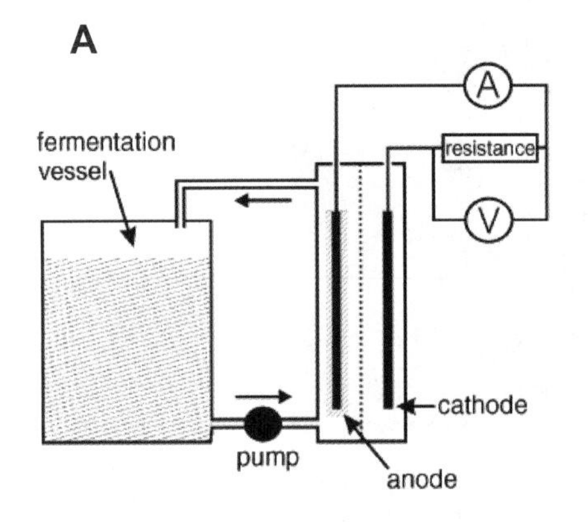

Fig. 6.23 (A) Schematic of a biohydrogen MFC (HMFC) where hydrogen is produced by fermentation in one vessel and is then used for electricity generation in a second reactor. [From *Schröder et al.* (2003), copyright Wiley-VCH Verlag GmbH & Co. KGaA, reproduced with permission.] (B) Photobiological HMFCs examined for electricity production by Rosenbaum *et al.* (2005). (C) Another reactor tested by this group for electricity production (*Logan et al.* 2006). (Reprinted with permission of the American Chemical Society.)

the bacterially-catalyzed systems. Because H_2 gas is produced in the anode chamber the anode potential is set by H_2 not by the respiratory enzymes of the bacteria (see Chapter 3). While bacteria catalyze the conversion of substrate into H_2, they are not the catalyst for electricity generation, and so an additional chemical catalyst is needed at the anode for H_2 oxidation.

In the first HMFC developed by this research group, H_2 was generated by bacteria in a separate fermentation vessel and then pumped into a fuel cell (Fig. 6.23A) (*Schröder et al.* 2003). The anode consisted of platinized woven graphite cloth that was modified with a conductive polymer (polyaniline); the cathode consisted of Pt wire, separated by a Nafion CEM. The reactor using *Escherichia coli* K12 to generate H_2 from glucose fermentation produced up to 1.5 mA/cm^2 based on the anode surface area, for a total power production of 6 W/m^2. In a subsequent study, it was found that performance could be improved using fluorinated polyanilines (*Niessen et al.* 2004). Besides glucose, starch and cellulose were also shown to be suitable substrates for biohydrogen production and electricity generation. The current and power densities achieved in the HMFCs can be substantially larger than those produced by conventional MFCs. However, the disadvantage of this system is that the substrate is fermented and not fully oxidized, and as a result most of the organic matter remains in solution. Typical fermentation conversion rates are ~2 moles of hydrogen produced per mole of glucose, with a maximum conversion of 4 mol/mol. This translates to approximately 17% conversion of substrate to hydrogen, with a maximum of 67% conversion for glucose.

To overcome the problem of residual organic matter in a HMFC, Rosenbaum *et al.* (2005) developed a photobiological HFMC (Fig. 6.23B). The non-sulfur purple bacterium *Rhodobacter sphaeroides* is known to be able to convert volatile organic acids, such as acetic acid produced as an end product in glucose fermentation, to H_2. Polyaniline was not used on the anode in tests with this microorganism as the performance of the reactor was degraded after only 5 to 10 h. Instead, a sulfur-based poly(3,4-ethylenedioxythiophene (PEDOT) material and Pt was applied to a graphite electrode. With *E. coli* used in a first-stage reactor to ferment glucose to H_2 and volatile acids, and *R. sphaeroides* used in a second-stage reactor with light to convert volatile acids to H_2, they found they could achieve 50% conversion of the residual organic acids produced by *E. coli* to H_2. Lactic acid was completely removed, while ethanol and acetic acid were removed but to a lesser extent. On an artificial mixture of volatile acids, a suspension of *R. sphaeroides* produced 180 mW/m^2. However, the reactor system was not optimized for volumetric power density. While power densities are appreciable, these photobiological systems have the limitation of needing sunlight for power generation. Photobiological conversion efficiencies are commonly thought to be a maximum of 10%, and with the system used by the authors they achieved an effective light conversion efficiency of 8.5%. Thus, while light-driven power generation is an interesting application, it is expected that the use of these photo-based systems would be limited in comparison to MFCs that can fully oxidize a wide range of organic substrates while producing continuous power without the need for sunlight.

Another alternative to the problem of residual organic matter is to find catalysts that can help to catalyze the oxidation of the remaining organic matter. Rosenbaum *et al.* (2006) found that using an anode electrode made with tungsten carbide, that they could achieve power generation using not only hydrogen gas, but also using formate. It may be possible to extend power generation via electrochemical catalysis of other fermentation end products to electricity as well.

6.9 Towards a scalable MFC architecture

Many different configurations, materials and operational modes have been used to produce power in MFCs with pure and mixed cultures, but there remains no proven design yet that is economical for scale-up. However, that situation will likely soon change for the following reasons. First, power densities are continually increasing with air cathodes. Second, high-cost (but high-performance) anode materials such as graphite plates are being replaced by low-cost materials such as graphite fiber brushes. Third, precious metal catalysts such as Pt are being replaced with little change in efficiency with non-precious metal catalysts using iron or cobalt. Fourth, reactor designs are being tested that allow closely spaced electrodes, a design that increases power output. Fifth, the reactors are being designed to maximize electrode packing (surface area per volume) so that power per reactor volume is increasing.

It appears likely that cost-effective reactor designs will be based on using graphite brush electrodes and tubular cathodes. The exact materials to be used with this architecture are still under development, but there are promising results based on bench-scale reactors that indicate this approach is feasible. The greatest challenge remaining is to devise the least expensive tube and a coating made of conductive and catalytically active material that will sustain power generation when exposed to the compounds present in a variety of wastewaters. When that is accomplished, tests and models will provide guidance on optimal loading and configurations of the electrodes in the reactor. These optimized MFCs can then be used in larger scale tests, with the first applications likely occurring in wastewater treatment plants. In Chapter 9, we explore how the reactor can fit into wastewater treatment plant process train, and what engineering challenges will remain for full-scale implementation of this exciting technology.

CHAPTER 7

Kinetics and Mass Transfer

7.1 Kinetic- or mass transfer-based models?

Current cannot be generated in an MFC at a rate greater than the rate bacteria can oxidize a substrate and transfer electrons to the surface. The current density, or current per area, is therefore related to the density of bacteria on a surface. Once bacteria cover a surface, they form a biofilm which can then grow (in theory) to have no limits, although in practice (with wastewater biofilms) the biofilm thickness reaches a few millimeters and then sloughs off due to dead bacteria at the solid interface. The thickness that can be achieved by an electrogenic biofilm, and the distance that bacteria can get from the surface and still use it as an electron acceptor, are not known. Indeed, the way an electrogenic biofilm develops is much different from that of a "typical" biofilm. The active and growing bacteria are near the supporting (conductive) surface in an electrogenic biofilm, not near the biofilm/fluid surface as in a typical aerobic biofilm in a wastewater treatment system such as a trickling filter.

As an electrogenic biofilm becomes thicker, we can expect that mass transfer to the biofilm could become limiting. This would result in the current density being limited not by the bacterial degradation of the organic matter, but by the diffusion or advection of substrate to the biofilm surface. There are certainly ample cases in the literature to suggest that biofilms can operate near the mass transfer limit of substrate transfer into the biofilm, depending on reactor conditions. For example, in trickling filters (TFs) the substrate flux from the falling film can approach the maximum rate that the bacteria can degrade the organic matter (*Logan* 1999; *Logan et al.* 1987). However, the analysis of the system is more complicated than just diffusion versus kinetics as the biofilm kinetics affects the rate of mass transfer.

In this chapter, boundaries are calculated on maximum current densities achievable by bacteria in biofilms by examining the upper limits to biofilm kinetics and comparing these to limits imposed by substrate mass transfer to the biofilm. With this information, we can calculate maximum power densities (power per surface area). We have already seen in Chapter 5 that there are limits on current and power densities imposed by internal resistance. Here, we examine the limits based on mass transfer calculations so that in a final analysis we can put limits on power density per area of the electrode that could be achieved in MFCs with a low internal resistance.

7.2 Boundaries on rate constants and bacterial characteristics

The concentration of bacteria, N_b, is often expressed as cell number concentration N (#/mL) when bacteria are grown as freely suspended (*i.e.*, non-flocculated) in culture. Alternatively, cell concentrations can be assessed in terms of mass concentration, X, on the basis of dry weight (DW), protein, or volatile suspended solids (VS). These values should be interchangeable by constants once they are known for a particular microorganism. However, the mass and the size of the cell changes over the growth cycle of the cells or when using different carbon sources or media. For our calculations here, we will assume a single cell size and that DW:protein:VS ratio is constant. With these assumptions, we can examine under what conditions microbial kinetics or mass transfer could limit power generation.

To begin developing a simple biofilm model based on kinetics, we first assume that bacterial growth is proportional to the concentration of bacteria, or

$$\frac{dX}{dt} = \mu X \tag{7-1}$$

where μ (1/t) is the specific growth rate that is a function of substrate concentration that can be described using Monod kinetics as

$$\mu = \frac{\mu_{\max} c}{K_c + c} \tag{7-2}$$

where μ_{max} (1/t) is the maximum growth rate, K_c is the half-saturation constant, and c is the substrate concentration. The rate that substrate is used by bacteria is assumed to be in proportion to the growth rate of the bacteria, or

$$\frac{dc}{dt} = -\frac{1}{Y_{X/c}} \frac{dX}{dt} \tag{7-3}$$

where $Y_{X/c}$ is the yield constant, here with units as specified of mass concentration of cells grown per mass concentration of substrate, or more simply as mass of cells per mass of substrate. Combining the above results, we have a rate expression for substrate concentration over time as

$$\frac{dc}{dt} = -\frac{X}{Y_{X/c}} \frac{\mu_{\max} c}{K_c + c} \tag{7-4}$$

This result shows that the rate of substrate utilization is not a simple function of substrate concentration as this rate changes depending on the specific value of c (Fig. 7.1). We further assume that the cell density in the biofilm is constant. That means that as the cells grow the biofilm will become thicker, but the packing density of the cells in the biofilm will not change. With this assumption, X = constant. This assumption of constant cell density is also valid for suspended cells if the cells produced over a time range of interest (dX/dt) is small compared to the initial cell concentration, X_0.

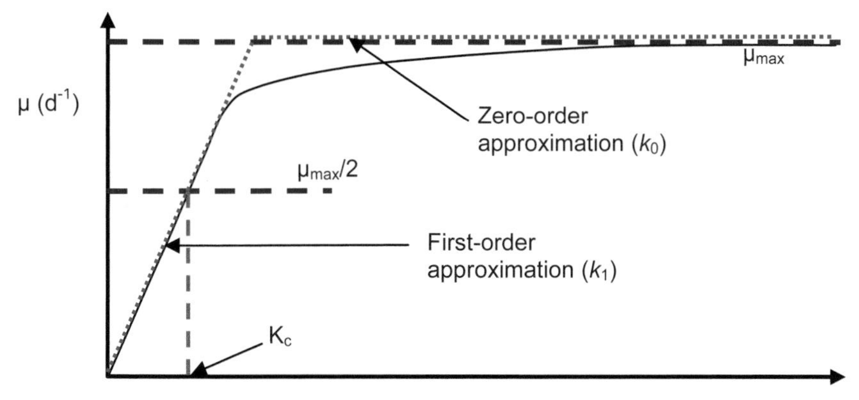

Fig. 7.1 Bacterial growth rate as a function of substrate concentration according to Monod kinetics (solid black line) with approximations for zero- and first-order kinetics (dotted lines).

We can examine two limits of the rate equation to make our calculations of biofilm kinetics easier: very high and very low substrate concentrations. When the substrate concentration is high (*i.e.,* non-limiting for growth), the bacteria grow at their maximum rate. Under these conditions, $c >> K_c$ and eq. 7-4 becomes equivalent to a zero-order rate constant, or

$$\frac{dc}{dt} = -\frac{\mu_{\max} X}{Y_{X/c}} = -k_0 \qquad \text{(zero-order approximation)} \qquad (7\text{-}5)$$

If the substrate concentration is small relative to the value of K_c, we can then make the assumption that $c << K_c$, and eq. 7-4 becomes

$$\frac{dc}{dt} = -\frac{X \mu_{\max}}{Y_{X/c} K_c} c = -k_1 c \qquad \text{(first-order approximation)} \qquad (7\text{-}6)$$

These two approximations form limits on the rate of substrate utilization by the biofilm when we consider a reasonable range of constants for cell growth kinetics and biofilm densities. To use these results, we must therefore examine what typical values we might obtain for these constants.

Mass of cells. To make conversions between bacterial size, volume, and projected area, we need to make some assumptions about a "typical" bacterium. First, let us consider the example of the well studied *Escherichia coli*. Statistics for this microbe include: 2 μm length × 0.8 μm diameter, cell volume of 1 μm^3, 1×10^{-12} g = 1 pg wet weight, 3×10^{-13} g = 0.3 pg dry weight (g-DW), 70% water and 30% solids, and 17% protein (56% of dry weight is protein) (*URL* 2007). These conditions are "average" for *E. coli* grown in rich nutrient media. Actual cell properties can change over a growth cycle and with the specific medium and growth conditions (batch or continuous).

There are no specific reports on cell weights for bacteria grown in MFCs. However, we can estimate cell weight for *Geobacter sulfurreducens* based on data from cells grown under other conditions. For *G. sulfurreducens* (0.5 μm diameter, 2.5 μm long) grown in syntrophic co-culture with hydrogen-oxidizing *Wollinella succinogenes* or *Desulfovibrio sulfuricans*, we obtain 1.5×10^{-12} g-DW/cell for both strains, for culture conditions of 180 mg-DW/L with 1.2×10^8 cell/ml (*Cord-Ruwisch et al.* 1998). This value is increased slightly to 1.7×10^{-12} g-DW/cell if we assume *G. sulfurreducens:W. succinogenes* ratios of 6.5:1 for cell DW (range 4–9) and 3:1 for cell number. Note that all these values are much larger than those obtained for *E. coli*, despite the similar sizes of the two bacteria.

Another approach is to estimate cell properties based on "typical" values. First, we assume a typical bacterium is rod shaped with a cell size of 0.5 μm × 2.5 μm. Assuming a cylinder, the cell has a projected surface area of $A_{cell} = 1.25\ \mu m^2$ (*i.e.*, what you would view through a microscope) and a volume ($v = \pi d^2 L/4$) of $v_{cell} = 0.491\ \mu m^3$. For an equivalent spherical cell, based on the volume, this is equivalent to a cell diameter of d_{cell} = 0.98 μm. Assuming that a cell has a density of 1.06 g/cm^3 and that it is 70% water (*i.e.*, it has a dry weight that is 30% of the mass), the mass of a single cell is

$$m_{cell} = \rho_{cell} v_{cell} f_{DW} = (1.06 \frac{g}{cm^3})\,(0.491 \mu m^3)\,(0.3)\,(\frac{cm}{10^4 \mu m})^3 = 1.6 \times 10^{-13}\ \frac{g\text{-}DW}{cell} \quad (7\text{-}7)$$

With this estimate, we get a dry weight of about half the typical dry weight of *E. coli*. Given the large variations in cell weights calculated above, and that *E. coli* are relatively large bacteria, we will assume here a value of 2×10^{-13} g-DW/cell in the calculations below. Often cell weights are measured in terms of protein concentration, so assuming a cell is 50% protein, this would be equivalent to 1.0×10^{-13} g-protein/cell.

Maximum growth rates. For cells growing exponentially (see eq. 7-1), the growth rate of a cell is related to its doubling time by $t_d = (\ln 2)/\mu$. Maximum doubling times for *E. coli* can reach 30 minutes ($\mu = 33\ d^{-1}$), but this is highly unusual for bacteria in nature as they have doubling times more on the order of one per day or longer. Doubling times for *G. sulfurreducens* in co-culture ranged from 6–8 h with *W. succinogenes* and 30 h with *Desulfovibrio desulfuricans* (*Cord-Ruwisch et al.* 1998). Let us assume here a maximum doubling time of 2 hours (12 doublings per day), or $\mu_{max} = 0.35\ h^{-1} = 8.3\ d^{-1}$.

Half-saturation constants. Kinetic constants for bacteria in MFCs have not been well studied. The data obtained so far shows that K_c values depend on the external resistance (load) on the circuit. Liu *et al.* (2005b) obtained values for K_c using acetate of 0.141 g/L (R_{ex} = 218 Ω, P_{max} = 661 mW/m^2), 0.043 g/L (R_{ex}=1000 Ω, P_{max} = 343 mW/m^2), and 0.009 g/L (R_{ex} = 5000 Ω, P_{max} = 86 mW/m^2). Thus, we see larger values of K_c with higher power densities and lower resistances. As we are interested in systems producing high power densities, we will use here an estimate for the half saturation constant for acetate of K_c = 0.2 g/L in the calculations below.

Biofilm porosities and densities. Biofilms are mostly (95–99%) water, with measured porosities that vary even over the thickness of the biofilm. For example, porosities ranged from 0.68 to 0.81 at the biofilm-water surface where cells are actively growing in aerobic biofilms, to 0.38–0.45 at the biofilm-support interface (*Zhang and Bishop* 1994b). The porosity and tortuosity of a biofilm reduces chemical diffusivities. Diffusion coefficients of chemicals in biofilms have been reported to be reduced by a factors of 0.5 and 0.8 in some experiments, with even wider ranges of 0.04 to 1.4 on other tests (*Logan* 1999). Solids concentrations leaving an activated sludge unit typically range from 0.5 to 1.5% (5 to 15 g/L), while those from a trickling filter fall in the range of 1–4% (*Metcalf & Eddy Inc.* 2003). Densities of wastewater biofilms insitu can reach 3–4 % or more (*Zhang and Bishop* 1994a). Values for biofilms produced in MFCs have so far not been reported. For our modeling purposes here, we will assume an estimate of the cell concentration in the MFC biofilm of $X = 30$ g/L or 3% solids.

Cell yields. The amount of biomass produced by bacteria per mass of substrate degraded is a highly variable number dependent on growth rate, conditions, the substrate, and point in the growth cycle when cells are analyzed. First, lets us consider "typical values" in the wastewater field, which are 0.4 g-COD-cell/g-COD-substrate for aerobic cell growth, and around one-tenth of that value for anaerobic growth. For BOD-based values, we can assume ~0.8 g-BOD_5/g-COD. For aerobic processes, a typical cell composition is $C_5H_7NO_2$ (113 g/mol), so there is 1.42 g-COD/g-cell, and for anaerobically grown cells we can use $C_{4.9}H_{9.4}O_{2.9}N$ (129 g/mol; 1.25 g-COD/g-cell), although these values vary (*Rittmann and McCarty* 2001).

Cell yields are not well-investigated for MFCs. One early report by Rabaey *et al.* (2003) indicated a range of 0.07 to 0.22 g-COD-cell/g-COD-substrate. More recently, Freguia *et al.* (2007) reported 0.304, 0.237, and −0.016 mol-C-biomass/mol-C-substrate obtained by mass balances around MFCs with acetate as a substrate and ferricyanide as the catholyte. They used a cell composition of $CH_{1.75}O_{0.52}N_{0.18}$ (1.38 g-COD-cell/g-cell), so these yields are equivalent to 0.31 and 0.24 g-COD-cell/g-COD-substrate (for the two positive yields). For glucose, values ranged from 0.10 to 0.51 g-COD/g-COD, depending on reactor acclimation and conditions.

In a pure culture study, Bond *et al.* (2002) measured 11.8 g-protein/mol-Ac (*Desulfuromonas acetoxidans*), or 24 g-cell/mol-Ac assuming protein is 50% of the cell dry weight in a two-chamber MFC. Using the same cell composition of $C_{4.9}H_{9.4}O_{2.9}N$, this is equivalent to 0.46 g-COD-cell/g-COD-substrate. This value can be compared to a range of values reported for *G. sulfurreducens* grown in co-culture. The cellular yield of this microbe with acetate (Ac) was 14 g-cell/mol-Ac (with *W. succinogenes*) and 2.3 g-cell/mol-Ac (with *Desulfovibrio desulfuricans*) (*Cord-Ruwisch et al.* 1998). Taking the average of these two values, we have $Y_{X/c} = 8.15$ g-cell/mol-Ac. Converting this to a COD basis, we have

$$Y_{X/c} = \left(\frac{8.15\ \text{g-cell}}{\text{mol-Ac}}\right)\left(\frac{\text{mol-Ac}}{60\ \text{g-Ac}}\right)\left(\frac{1.25\ \text{g-COD-cell}}{\text{g-cell}}\right)\left(\frac{\text{g-Ac}}{1.08\ \text{g-COD-Ac}}\right) \quad (7\text{-}8)$$

$$Y_{X/c} = 0.16\,\frac{\text{g-COD-cell}}{\text{g-COD-Ac}} \quad (7\text{-}9)$$

A yield of 0.16 g-COD/g-COD is within the wide range reported above by Rabaey *et al.* (2003) for MFCs but lower than that obtained by Freguia *et al.* (2007). We will use a value of 0.16 g-COD/g-cell as a "typical" cell yield in an MFC even though we recognize this value is not well defined and is likely to vary from system to system.

This result based on 0.16 g-COD/g-COD is equivalent to 6.8×10^{11} cell/g-Ac, or 4.1×10^{13} cell/mol-Ac (assuming 1.25 g-COD/g-cell and 2×10^{-13} g-DW/cell). This value is much higher than a report for iron-reducing bacteria grown under anaerobic conditions where cell yields of *G. sulfurreducens* on acetate under anoxic conditions (with fumarate as an electron acceptor) was 5.7×10^{10} cells/mol-Ac, versus 5.9×10^{10} and 6.3×10^{10} cells/mol-Ac in cultures with 5 and 10% O_2, respectively (*Lin et al.* 2004). In contrast, the value calculated above for the MFC study by Bond and Lovley (2003) implies 1.2×10^{14} cells/mol-Ac. With values ranging from 10^{10} to 10^{14} cells/mole, further research is needed in this area to better determine values appropriate for cell production in MFCs.

Example 7.1

For the yield calculated in eq. 7-9, (a) what is the implied C_E if the only fate of acetate is biomass production and generation of electricity? (b) Bond and Lovley (2003) reported a maximum C_E= 96.8% for *Geobacter sulfurreducens* using acetate, and Chaudhuri and Lovley (2003) found C_E = 83% for *Rhodoferax ferrireducens* using glucose. What does that imply about the cell yield?

(a) The yield in the above equations is dimensionless because both the numerator and denominator are expressed on the basis of COD. Thus, the C_E represents flow of electrons into the circuit (substrate used for cell respiration) while the Yield is a biomass efficiency (efficiency of converting substrate into biomass). If substrate only goes to biomass and electricity, then $Y_{X/c} + C_E = 1$, or

$$C_E = 1 - Y_{X/c} = (1 - 0.16) = 0.84 \qquad (7\text{-}10)$$

The actual CE will be lower than this as substrate is used by bacteria for other purposes (*i.e.*, for aerobic growth or growth using alternate electron acceptors). Thus, we can write in general that the fraction of substrate (or electrons or COD) that go to other processes, φ, is defined as (*Logan et al.* 2006)

$$\varphi = 1 - C_E - Y_{X/c} \qquad (7\text{-}11)$$

(b) Based on the above analysis, the value for *R. ferrireducens* seems reasonable as this implies a cell yield of $Y_{X/c}$ = 0.17 g-COD-cell/g-COD-substrate, which is quite similar to our estimate above. However, the result for *G. sulfurreducens* would be extraordinarily low as it implies a yield of $Y_{X/c}$ = 0.032 g-COD/g-COD. This low yield would imply that *G. sulfurreducens* is a poor competitor for making cell biomass from substrate, which is unexpected given the relative importance of *Geobacter* spp. for iron reduction in natural environments.

7.3 Maximum power from a monolayer of bacteria

If we assume that bacteria must directly contact a surface in order to transfer electrons to it, then we can make our first estimate of maximum power production for a flat surface.

Assuming cells fit completely on the surface (*i.e.,* the packing efficiency is 100% and there is no exposed surface area, the number of cells on the surface is $X = A/A_{cell}$. However, even a surface apparently well-covered by bacteria will not be completely packed. For example, the surface shown in Fig. 7.2 looks covered by bacteria, but a count of the microspheres normalized to the surface area indicated that only 30% of the surface was covered. In bacterial adhesion experiments, a surface acts as if it is completely covered (referred to as the "jamming" limit) when as little as 3–5% of the sites are occupied by bacteria (*Vadillo-Rodriguez and Logan* 2006). Even tightly spaced spherical particles will only cover 54% of the surface due to packing limits.

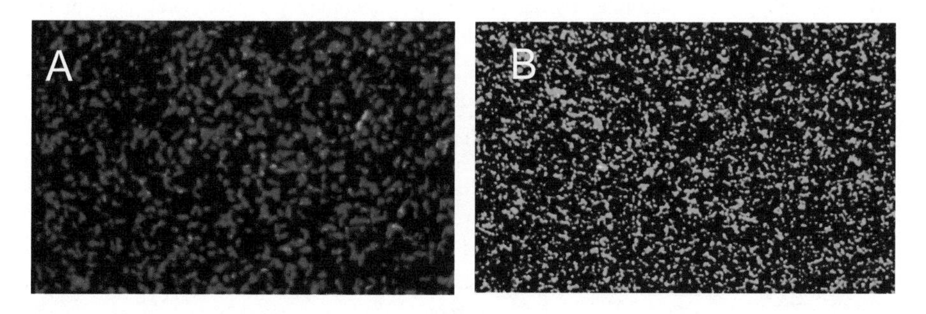

Fig. 7.2 (A) *Escherichia coli* D21 and (B) fluorescent latex microsphere adhesion to a surface. The coverage of the surface is 30% for *E. coli.* [From Pardi (2002), with permission.]

The current that can be generated based on the rate of substrate utilization, *dc/dt*, for *X* cells, and the measured Coulombic efficiency, is

$$I = -\frac{dc}{dt} b_{es} F C_E \tag{7-12}$$

where b_{es} is the number of moles of electrons per mole of substrate (COD in mg/L). Assuming the maximum rate of substrate removal, or $\mu = \mu_{max}$, the current is

$$I = \frac{\mu_{max} X}{Y_{X/c}} b_{es} F C_E \tag{7-13}$$

Power is then calculated based on $P = IE$, resulting in

$$P = \frac{\mu_{max} X b_{es} F C_E E}{Y_{X/c}} \tag{7-14}$$

For the maximum power we can use a cell voltage equal to the open-circuit voltage (*i.e.,* $E = OCV$), but in practice the voltage produced will be lower.

Example 7.2

Calculate the maximum power based on surface area using the assumptions above for average conditions in an MFC with acetate as a substrate, and an *OCV* of 0.8 V.

Assuming the surface area per cell calculated above of A_{cell} = 1.25 μm^2, along with complete surface coverage, the cell density per square meter of flat surface would be

$$X = \frac{A_{An}}{A_{cell}} = \frac{1\,\text{m}^2}{1.25\,\mu\text{m}^2/\text{cell}}\left(\frac{10^6\,\mu\text{m}}{\text{m}}\right)^2 = 8\times10^{11}\,\frac{\text{cells}}{\text{m}^2}$$

We assume a high growth rate of μ_{max} = 8.3 d^{-1} (12 doublings per day), a cell yield of $Y_{X/c}$ = 4.1×10^{13} cells/mol-Ac, and $b_{e\,s}$ = 8 mol e$^-$/mol-Ac. The maximum power with an *OCV* = 0.8 V, is

$$P = \frac{\mu_{\max}\, X\, b_{es}\, F\, C_E\, E}{Y_{X/c}} = \frac{(8.3\,\text{d}^{-1})(8\times10^{11}\,\text{cell})}{(4.1\times10^{13}\,\text{cell/mol-Ac})}\left(\frac{8\text{e}^-}{\text{mol-Ac}}\right)\left(\frac{96500\,\text{C}}{\text{mol e}^-}\right)$$

$$\times(0.84)\,(0.8\,\text{V})\frac{\text{d}}{86{,}400\,\text{s}} = 0.97\,\frac{\text{W}}{\text{m}^2}$$

The result shown in the above example is a power density that has been exceeded in many studies, and thus this calculation provides good evidence that cells form more than a monolayer on a complete flat surface. For two layers of cells, we could simply double this number, and for three layers, triple it, and so forth. Cells cannot completely fill a surface, and dead cells may occupy some part of the surface, which would decrease the above estimate. More importantly, it must be recognized that most electrode surfaces are not completely flat and thus it is possible that we have underestimated the surface area and thus the maximum cell density. The result shown is slightly different from that obtained by Liu and Logan (2004) (2.2 W/m^2) due to their use of a different yield (Y = 3.6 × 10^{13} cell/mol-Ac), smaller cells, and the lack of inclusion of the Coulombic efficiency in the calculation versus C_E = 0.84 used here.

7.4 Maximum rate of mass transfer to a biofilm

As bacteria colonize a surface and form a thick biofilm, the rate of substrate consumption possible by the biofilm can eventually exceed mass transfer (by diffusion) to that surface. Here we consider the possibility that substrate flux to the biofilm limits power generation, *i.e.*, that the biofilm is able to discharge all possible current to the surface. This situation could be expected to arise at high current densities, as shown in Fig. 4.3 (see section 3 of the figure) where current rapidly decreases due to mass transfer limitations.

To model the biofilm, we take the approach that it is a type of catalyst layer, where the flux to the biofilm is described by a mass transfer coefficient, $\boldsymbol{k}_w$, as

$$J_b = \boldsymbol{k}_w\,(c - c_{b0}) \tag{7-15}$$

where c is the bulk substrate concentration near the anode, and c_{b0} is the concentration of substrate at the biofilm surface. Correlations for the mass transfer coefficient are available for different fluid mechanical environments, and are usually expressed in terms of a stagnant film thickness, δ_s (although stagnant films don't exist in practice), and the chemical diffusion constant in water, D_{Cw}, as $\boldsymbol{k}_w = D_{Cw}/\delta_s$ (*Logan* 1999). The maximum rate of mass transfer can be obtained when c_{b0}=0, but this situation can't actually occur as there must be some finite concentration of substrate in the biofilm. A more extensive discussion of biofilm modeling can be found in more specialized textbooks (*Logan* 1999; *Rittmann and McCarty* 2001).

Let us take the simplifying approach that there is no external mass transfer limitation, as this approach will yield an upper limit on the chemical flux into a biofilm that is not limited by mass transfer outside the biofilm. To do that, we assume that the concentration at the biofilm surface is the bulk substrate concentration.

First-order biofilm kinetics. If we assume first-order biofilm kinetics, with a rate constant k_1, then the maximum flux of substrate into the biofilm can be calculated (*Logan* 1999) as

$$J_b = k_1\, \delta_b\, c\, \frac{\tanh(B_1)}{B_1} \qquad \text{(first-order biofilm kinetics)} \qquad (7\text{-}16)$$

where δ_b is the biofilm thickness, and B_1 is a dimensionless constant defined as

$$B_1 = \left(\frac{k_1\, \delta_b^2}{D_{Cb}} \right)^{1/2} \qquad (7\text{-}17)$$

where D_{Cb} is the diffusion coefficient of the substrate in the biofilm (which is less than or equal to its value in water). If the rate of reaction is very fast relative to diffusion, then B_1 becomes large and tanh $(B_1) \rightarrow 1$. For this condition, we can simplify the above equations and write:

$$J_{C,z} = (k_1\, D)^{1/2}\, c \qquad (\textit{very fast reaction}, \text{first-order kinetics}) \qquad (7\text{-}18)$$

The rate of substrate flux to the surface can never be larger than this flux as external mass transfer limitations could reduce the concentration of substrate at the surface to be less than *c*.

To calculate the maximum current produced by this substrate flux, we convert this flux to the electron flux (current density) using the same approach as in eq. 7-12, to obtain for the maximum current and power

$$I_{\max} = (k_1\, D)^{1/2}\, b_e\, F\, C_E\, c \qquad (7\text{-}19)$$

$$P_{\max} = (k_1\, D)^{1/2}\, b_e\, F\, C_E\, E\, c \qquad (7\text{-}20)$$

Example 7.3

Calculate the maximum power density (W/m^2) for a biofilm assuming first-order kinetics and using the above values for typical biofilms, assuming a diffusion constant of $D_b = D = 0.88 \times 10^{-5}$ cm^2/s, an OCV of 0.8 V, and $C_E = 0.84$. Assume a substrate concentration of $c = K_c = 0.2$ g/L so that the specific growth rates of the cells remains in the first-order region (see Fig. 7.1).

First, we need to calculate the first-order rate constant, assuming $X = 30$ g/L, $\mu_{max} = 8.3$ d^{-1}, a cell yield of $Y_{X/c} = 8.15$ g-cell/mol-Ac, and $b_e = 8$ e$^-$/mol-Ac

$$k_1 = \frac{X\,\mu_{\max}}{Y_{X/c}\,K_c} = \frac{(30\,\text{g-cell/L})\,(8.3\,\text{d}^{-1})}{(8.15\,\text{g-cell/mol-Ac})(0.2\,\text{g-Ac/L})}\left(\frac{60\,\text{g-Ac}}{\text{mol-Ac}}\right)\frac{\text{d}}{86{,}400\,\text{s}} = 0.11\,\text{s}^{-1}$$

Next, we calculate the value for B_1 and check to see if this is a "fast reaction".

$$B_1 = \left(\frac{k_1\,\delta_b^2}{D_{Cb}}\right)^{1/2} = \left(\frac{(0.11\,\text{s}^{-1})\,(0.1\,\text{cm})^2}{(0.88\times10^{-5}\,\text{cm}^2/\text{s})}\right)^{1/2} = 11.2$$

Because tanh (11.2) = 1, we can assume this is a fast reaction. For acetate, $b_{es} = 8$ e$^-$/mol-Ac

$$P_{\max} = (k_1\,D_{Cb})^{1/2}\,b_{es}\,F\,C_E\,E\,c$$

$$P_{\max} = \left[(0.11\,\text{s}^{-1})(0.88\times10^{-5}\,\text{cm}^2/\text{s})\right]^{1/2}\left(\frac{8\,\text{mol-e}^-}{\text{mol-Ac}}\right)\left(\frac{96500\,\text{C}}{\text{mol-e}^-}\right)(0.84)\,(0.8\,\text{V})$$

$$\times\left(\frac{0.2\text{g-Ac}}{\text{L}}\right)\left(\frac{\text{mol-Ac}}{60\,\text{g-Ac}}\right)\frac{\text{L}}{10^3\,\text{cm}^3}\,\frac{10^4\,\text{cm}^2}{\text{m}^2} = 17\,\frac{\text{W}}{\text{m}^2}$$

Note that using the thermodynamic potential of $E^{0/}{}_{\text{cell}} = 1.105$ V (which cannot be reached) and 100% C_E, we would obtain $P_{max} = 28$ W/m^2. Also, we have used here a value of $\mu = ½\,\mu_{\max}$, so the maximum power density could still be larger than the above value, but the kinetics would no longer be first-order.

Zero-order biofilm kinetics. For this case, we have two different solutions for the substrate flux depending on the thickness of the biofilm. For a "shallow" biofilm, meaning that the substrate completely penetrates the biofilm and reaches the support surface (what is needed for electrochemical activity at the electrode surface), the flux is

$$J_b = k_0\,\delta_b \qquad \text{(zero-order, shallow biofilm, } B_0 \geq 1) \tag{7-21}$$

This solution is only valid if the substrate is not used up before it reaches the bottom of the biofilm. Thus, the above equation is only valid for $B_0 > 1$ where B_0 is defined as

$$B_0 = \left(\frac{2\,D_{Cb}\,c}{k_0\,\delta_b^2}\right)^{1/2} \tag{7-22}$$

For a "deep" biofilm, *i.e.*, for conditions where the substrate is completely consumed before it reaches the biofilm/support surface, the substrate flux is

$$J_{C,z} = (2k_0 D_{Cb} c)^{1/2} \quad \text{(zero-order, deep biofilm)} \qquad (7\text{-}23)$$

While the flux predicted for the deep biofilm may be larger, we have the condition that substrate would not reach the electrode. While it may be possible that cells distant from the electrode could conduct electrons to it using nanowires or mediators, we shall restrict our calculations here for the first case of a shallow biofilm where the biofilm thickness is the maximum possible based on B_0. For this case, we have the current and power densities for zero order kinetics of

$$I_{\max} = k_0 \delta_b b_{es} F C_E \qquad (7\text{-}24)$$

$$P_{\max} = k_0 \delta_b b_{es} F C_E E \qquad (7\text{-}25)$$

Example 7.4

For the case of zero-order kinetics, calculate the maximum power density (W/m^2) for a biofilm using typical biofilm constants, $D_{Cb} = 0.88 \times 10^{-5}$ cm^2/s, and an $OCV = 0.8$ V. Here we will use a substrate concentration of 1 g/L of acetate so that we are in the zero-order kinetic region.

The zero-order rate constant, assuming $X = 30$g/L, $\mu_{max} = 8.3$ d^{-1}, a cell yield of $Y_{X/c} = 8.15$ g-cell/mol-Ac, is

$$k_0 = \frac{X \mu_{\max}}{Y_{X/c}} = \frac{(30\,\text{g-cell/L})(8.3\,\text{d}^{-1})}{(8.15\,\text{g-cell/mol-Ac})}\left(\frac{60\,\text{g-Ac}}{\text{mol-Ac}}\right)\frac{\text{d}}{86{,}400\,\text{s}} = 0.021\frac{\text{g}}{\text{L s}}$$

To calculate the biofilm thickness, we rearrange the equation with $B_0 = 1$, and solve it to obtain

$$\delta_b = \left(\frac{2 D_{Cb} c}{k_0}\right)^{1/2} = \left(\frac{2(0.88 \times 10^{-5}\text{cm}^2/\text{s})(1\,\text{g/L})}{(0.021\,\text{g/L-s})}\right)^{1/2} = 0.029\,\text{cm}$$

The power density is therefore calculated as

$$P_{\max} = k_0 \delta_b b_{es} F C_E E$$

$$P_{\max} = (0.021\frac{\text{g}}{\text{L s}})(0.029\,\text{cm})\left(\frac{\text{mol-Ac}}{60\,\text{g-Ac}}\right)\left(\frac{8\,\text{mol-e}^-}{\text{mol-Ac}}\right)\left(\frac{96500\,\text{C}}{\text{mole}^-}\right)(0.84)(0.8\,\text{V})$$

$$\times \frac{\text{L}}{10^3\,\text{cm}^3}\frac{10^4\,\text{cm}^2}{\text{m}^2} = 53\frac{\text{W}}{\text{m}^2}$$

The above calculations demonstrate that much higher power densities are possible before mass transfer to the biofilm will limit power generation. Note that these calculations do not consider the effect of the external mass transfer coefficient, *i.e.*, conditions where substrate flux to the biofilm reduces the possible substrate concentration at the surface. However, these calculations suggest that factors other than maximizing substrate mass transport to the biofilm should be emphasized for now, in agreement with polarization curves that suggest mass transfer limitations are not affecting power generation (*e.g.*, Fig. 4.2).

Calculations based on energy. The above analysis is all based on current transfer to the surface. However, another analysis approach has been to use the energy of the substrate in the flux calculation. Energy calculations are most useful when trying to determine the efficiency of the system for energy recovery. However, they do imply assumptions that cannot be met in an MFC, as shown in the example below.

Example 7.5

Repeat the maximum power density calculation done in Example 7.3, but this time use the heat of combustion for acetate (870.28 kJ/mol-Ac = 14.5 kJ/g-Ac) instead of the current-voltage approach.

For this case, we first calculate the flux as

$$J_b = (k_1 D_{Cb})^{1/2} c = \left[(0.11\,\mathrm{s}^{-1})(0.88\times10^{-5}\,\mathrm{cm}^2/\mathrm{s})\right]^{1/2} \times (\frac{0.2\,\mathrm{g\text{-}Ac}}{\mathrm{L}})\ \frac{\mathrm{L}}{10^3\,\mathrm{cm}^3}$$

$$J_b = 1.97\times10^{-7}\,\frac{\mathrm{g}}{\mathrm{cm}^2\,\mathrm{s}}$$

$$P_{max} = J_b \Delta H = (1.97\times10^{-7}\,\frac{\mathrm{g}}{\mathrm{cm}^2\,\mathrm{s}})(\frac{14.5\,\mathrm{kJ}}{\mathrm{g}})\ \frac{10^4\,\mathrm{cm}^2}{\mathrm{m}^2}\ \frac{3600\,\mathrm{s}}{\mathrm{h}}\ \frac{10^3\,\mathrm{W\,h}}{3600\,\mathrm{kJ}}\ \mathrm{P}_{max} = 28\,\frac{\mathrm{W}}{\mathrm{m}^2}$$

Note that this is larger than our result above based on using the *OCV* and C_E, but exactly the same as that obtained if we assume C_E = 100% and we use the maximum cell potential of $E^{0/}_{cell}$ = 1.105 V.

7.5 Mass transfer per reactor volume

The above analysis demonstrates that we are still below maximum power densities based on surface areas, but the design of MFCs will require systems that are efficient at power generation on a volumetric basis. Ultimately, the upper limit for power generation by bacteria is that of a single cell. If we had no mass transfer limitations, bacteria grew at the maximum rate, and there was no space between bacteria, then we would have the maximum volumetric power density. We can calculate this upper limit using assumptions about cell growth rates and cell size, as shown in the example below

---

Example 7.6

Calculate the maximum power density that can be achieved by a bacterium, using data typical of our previous calculations but assuming a doubling time of 30 minutes.

We can use eq. 7-5 to calculate the intrinsic rate of substrate removal, but first we need a value for the cell density. In the limiting case, this is just the density of a single cell, which we can calculate as:

$$X = \frac{m_{cell}}{v_{cell}} = \frac{(2\times10^{-13}\,\text{g})}{(0.491\,\mu\text{m}^3)}\frac{10^{12}\,\mu\text{m}^3}{\text{cm}^3} = 0.4\,\frac{\text{g-cell}}{\text{cm}^3}$$

Assuming a doubling time of 0.5 h, we can calculate the maximum growth rate and then the maximum substrate consumption rate (equal to minus the removal rate) as

$$\mu = \frac{\ln 2}{0.5\,\text{h}} = 1.39\,\text{h}^{-1}$$

$$\frac{dc}{dt} = \frac{-\mu_{\max} X}{Y_{X/c}} = \frac{-(1.39\,\text{h}^{-1})(0.4\,\text{g-cell/cm}^3)}{(8.15\,\text{g-cell/mol-Ac})} = -0.068\,\frac{\text{mol-Ac}}{\text{h-cm}^3}$$

Now, using eq. 7-14, we can calculate the power density as

$$P_v = -\frac{dc}{dt} b_{es}\, F\, C_E\, E = (0.68\,\frac{\text{mol-Ac}}{\text{h-cm}^3})(\frac{8\,\text{mol}\,\text{e}^-}{\text{mol-Ac}})(\frac{96500\,\text{C}}{\text{mol}\,\text{e}^-})(1)(1.105\,\text{V})\frac{10^6\,\text{cm}^3}{\text{m}^3}\frac{1\,\text{h}}{3600\,\text{s}}$$

$$P_v = 1.6\times10^8\,\frac{\text{W}}{\text{m}^3} = 160\,\frac{\text{MW}}{\text{m}^3}$$

The above example shows that a bacterium growing very fast actually is quite a power generator! However, such a power density is unrealistic in terms of the power that could be produced by bacteria in a biofilm in a reactor, even if we do not consider mass transfer limitations. If we assume cells grow in a biofilm that is 80% porous (20% filled with bacteria) and that the reactor porosity must be 95% (only 5% filled with bacteria, the above result would be reduced to 1.6 MW/m^3. This is still an unreasonably large number, and it is likely not to be achieved in a real due to mass transfer limitations and ohmic resistance.

An alternate approach for estimating an upper limit for volumetric power density is to consider the specific surface areas (area per volume) of electrodes that could be possible in an MFC. Most MFCs in the laboratory have used less than ~100 m^2/m^3, except for recent studies using RVC or graphite fiber electrodes. Specific surface areas typical of trickling filters used for wastewater treatment are ~10^2 m^2/m^3, but it is likely these can be increased for electrogenic reactors. Biofilms grow on electrode surfaces from the surface outward (*i.e.*, cells growing on the surface push out new cells), but in a trickling filters they grow at the water–biofilm interface, creating thick biofilms of dead material. Graphite fiber brush electrodes can now produce specific surface areas for the anode of 10^4 m^2/m^3, thus producing essentially unlimited growth for the bacteria, but even with these systems mass transfer of substrate into the brush electrode (or protons out of the electrode) could limit maximum power densities.

If we consider our previous calculations of mass transfer limited power generation in the range of 17 to 53 W/m^2, and a reasonable limit of 10^3 m^2/m^3, this implies an upper limit of 17 to 53 kW/m^3 if the system is limited by the surface area of the anode. However, it is likely that the cathode surface area will limit performance. Assuming cathode surface areas of 10^2 to 10^3 m^2/m^3, this would place the limit at ~2 to 53 kW/m^3.

So far, MFCs have yet to exceed 0.115 kW/m^3, so there is much room for improvement. It is estimated that if power could be produced above 0.4 kW/m^3, that MFCs could compete with anaerobic digestion processes. We can expect that as MFCs further develop we may soon start to achieve these upper power densities, at least in industrial applications where microbial communities or pure culture conditions can be used and carefully controlled.

CHAPTER 8

MECs for Hydrogen Production

8.1 Principle of operation

The evolution of hydrogen in an MFC-like reactor is made possible by release of electrons by exoelectrogenic bacteria. We refer to this process as *electrohydrogenesis*, as electrons are released by bacteria but they are combined with protons to form hydrogen gas as a product, not electricity as in an MFC. The reactor used for this process has been given different names, including: a bioectrochemically assisted microbial reactor (BEAMR) (*Ditzig et al.* 2007; *Liu et al.* 2005c; *Logan and Grot* 2005); and a biocatalyzed electrolysis cell (BEC) as it is a process dependent on the biocatalyzed electrolysis of organic matter (*Logan et al.* 2006; *Rozendal and Buisman* 2005; *Rozendal et al.* 2006a). If we follow the simple nomenclature used for the MFC, it should be called a "microbial electrolysis cell" (MEC) as an MFC produces electricity like a fuel cell, and an MEC produces hydrogen like an electrolysis cell. We follow here the convention in nomenclature developed for MFCs for naming the process, microorganisms, and reactor (electrogenesis, exoelectrogens, and MFC) using the terms electrohydrogenesis, exoelectrogens, and MEC.

The terms "electrochemically" and "assisted" are used in these process names because additional voltage must be added to the circuit. In an MFC an anode potential can approach the theoretical limit of $E_{An} = -0.3$ V of the substrate (*e.g.*, acetate, Table 3.1, 298 K, pH = 7). With oxygen, the cathode potential in an MFC is ~0.2 V (vs. NHE), achieving an overall a cell voltage approaching ~0.5 V [0.2 V – (−0.3 V) = 0.5 V). If we wish to form hydrogen at the cathode, however, we must remove the oxygen and overcome a cathode potential at pH = 7 and 298 °C of

$$E_{Cat} = E^0 - \frac{RT}{nF}\ln\frac{H_2}{[H^+]^2} = 0 - \frac{(8.31\,\text{J/mol K})\,(298.15\,\text{K})}{(2)\,(9.65\times10^4\,\text{C/mol})}\ln\frac{1}{[10^{-7}]^2}$$
$$= -0.414\ \text{V} \qquad (8\text{-}1)$$

Therefore, the calculated cell voltage for a system that could produce hydrogen at the cathode is

$$E_{emf} = E_{Cat} - E_{An} = (-0.414\,\text{V}) - (-0.3\,\text{V}) = -0.114\,\text{V} \tag{8-2}$$

The cell voltage is negative, so the reaction is not spontaneous. This helps explain why bacteria that produce acetate and H_2 cannot further convert the remaining acetate to H_2: the reaction becomes endothermic (*i.e.,* requires energy) once the acetate is formed at appreciable H_2 concentrations. However, H_2 gas can theoretically be generated in a MEC by adding a voltage of >0.114 V (*Liu et al.* 2005c; *Logan and Grot* 2005). Thus, the combination of the "bio" (bacterial electrolysis of organic matter) and "electrochemically assisted" (adding voltage) components in a single reactor allows us to produce H_2 gas in the MEC process.

The voltage needed to achieve H_2 gas production in a MEC can be applied using an MFC or any power source. If higher potentials could be obtained at the anode with

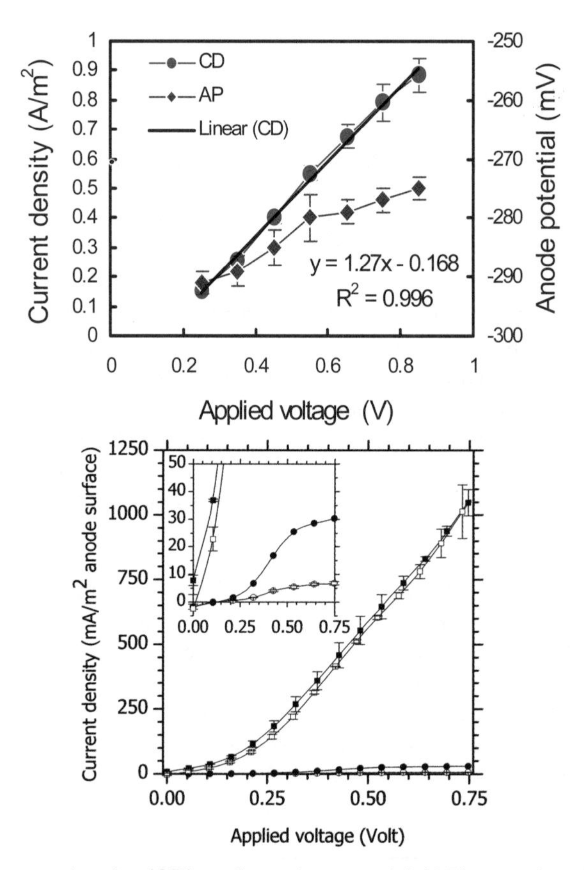

Figure 8.1 Current density (CD) and anode potential (AP) as a function of applied voltage in a two-chambered BEAMR reactor. (A) Results obtained by Liu *et al.* (2005c) using a two-bottle system, modified to show a linear regression of the current density. (Reprinted with permission from the American Chemical Society.) (B) System used by Rozendal *et al.* (2006a). (Reprinted with permission, International Journal of Hydrogen Energy.) (Error bars ±S.D.)

bacteria degrading more energetic substrates than acetate, it might be possible to generate hydrogen spontaneously by producing current without intermediates such as acetate. For example, for glucose at 298 K and pH = 7, theoretically a potential of $E^{0/} = -0.428$ V could be produced (Table 3.1). The cell emf for this case would be positive, or $E_{emf} = (-0.414 \text{ V}) - (-0.428 \text{ V}) = 0.014$ V, indicating that theoretically for glucose this reaction could be spontaneous. However, if glucose degradation proceeded via fermentation to acetate, then a spontaneous reaction would not be possible.

While 0.114 V *in theory* is needed with acetate as a substrate, in practice larger voltages must be applied due to overpotential at the cathode. Experiments have shown that *in practice* ~0.25 V or more must be applied to the circuit to obtain reasonable current densities, and thus useful generation rates of hydrogen (Fig. 8.1). Note that there is a nearly linear increase in the current density with the applied voltage. Analysis of this response can work like the analysis of a fuel cell using a Tafel plot, and thus the slope and intercept can be used to analyze the performance of the system. For the tests by Liu *et al.* (2005c) we see that the current density increases as a factor of 1.27 A/m^2 per volt, or 0.003 A/V. Using $R = E/I$, this is a resistance of 330 Ω. The disc-reactor system developed by Rozendal *et al.* (2006a) has a slope of 1.78 A/m^2 or 0.08 A/V, and therefore a lower resistance of 12.5 Ω. The x-intercept is a measure of the overpotential at both electrodes plus that needed to initiate hydrogen production. For these two cases, this potential is slightly less at 0.13 V for tests by Liu *et al.* (2005c) compared to 0.17 V for Rozendal *et al.* (2006a) (Fig. 8.1).

While we can increase current density by increasing voltage, the greater the applied voltage the larger the amount of energy input into the reactor (see below). Adding 0.25 V, however, is still substantially less than the voltage needed for water electrolysis which in practice is typically 1.8 to 2.0 V. This is because this electrolysis of organic matter (*Rozendal et al.* 2006a), or the splitting of organic matter into protons and electrons by bacteria, can be a thermodynamically favorable process (*i.e.*, exothermic) with a suitable electron acceptor (or via fermentation, depending on the substrate). In contrast, water splitting is always endothermic. The production of H_2 in a MEC process, rather than electricity generation in an MFC, is still an endothermic process overall, and thus energy needs to be added to the reactor in order to make hydrogen evolution possible.

8.2 MEC systems

The invention of the MEC system is a recent development, so there are not many systems that have been tested and reported in the literature. So far, there have been reports of four different systems used for H_2 production with the MEC process: three using acetate as a substrate; and one using domestic wastewater. These systems were typically two-chamber reactors with the anode and cathode chamber separated by a proton exchange membrane (Nafion 117), although one system was modified to contain a gas diffusion electrode (*Rozendal et al.* 2007). Voltage is added into the circuit using a power source, and the current is determined by monitoring the voltage across a low ohm-resistor (~10 Ω) placed in series using a multimeter. The use of a resistor results in a small loss of power (Fig. 8.2A), but some power sources can automatically correct for this loss (*Rozendal et al.* 2007).

The first system reported for H_2 evolution was developed by Liu *et al.* (2005c) using a two-bottle (310 mL each) reactor, with the chambers filled to 200 mL and separated by a small cation exchange membrane (CEM; Nafion) held in a tube between the reactors

(Fig. 8.2B). The anode was plain carbon cloth and the cathode was made of carbon paper with 0.5 mg-Pt/cm^2 (12 cm^2 for each electrode). This system was the same design used by Oh *et al.* (2004) in MFC tests, where the electrodes were spaced 15 cm apart. Using acetate, it was found that, on average, 2.9 mol-H_2 were produced per mole of acetate, compared to a maximum theoretical yield of 4 mol/mol. This required an energy input (voltage added to the circuit) equivalent to 0.5 mol of H_2 (lower heating value) at an applied voltage of 0.25 V. The recovery of electrons from acetate as current was high in this system, with a hydrogen Coulombic efficiency that ranged from C_E = 60% to 78% depending on the applied voltage. The recovery of these electrons from the circuit as hydrogen was greater than 90%. The internal resistance for this reactor is high, but it used a small CEM which limited the back diffusion of hydrogen from the cathode chamber.

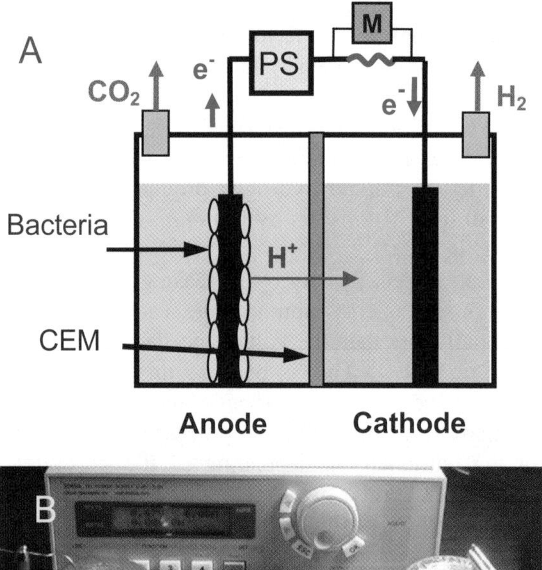

Fig. 8.2 (A) Generalized schematic and (B) photograph of the BEAMR process showing two chambers separated by a CEM with the voltage set using a power supply (PS) and a multimeter (M) used to monitor the voltage across a low ohm resistor Gas is sampled (or released) via the ports on the top of the reactors. [Adapted from *Liu et al.* (2005c) with permission from the American Chemical Society.]

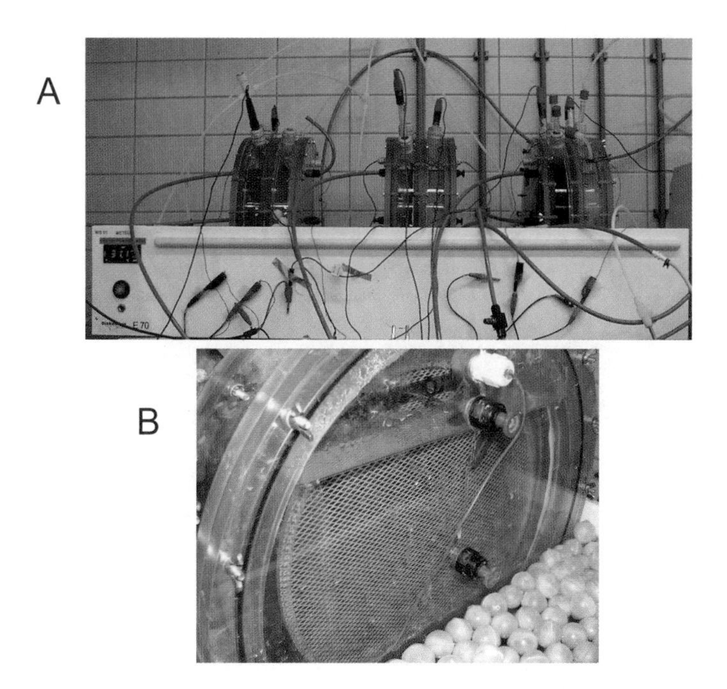

Fig. 8.3 Reactors used for acetate conversion to hydrogen using biocatalyzed electrolysis of organic matter by Rozendal *et al.* (2006a). (A) Three-reactor setup in a water bath; (B) close-up of the cathode. [Photographs by R. Rozendal.]

A second reactor was examined by Liu *et al.* (2005c) based on the tubular-reactor design of Liu and Logan (2004) (see Fig. 6.1). This single-chamber reactor was converted into a two-chamber system by inserting a CEM into the middle of a cube-type MFC that had a 4-cm-long cylindrical chamber and a 3 cm diameter, with both sides sealed off from air. Gas was released from the cathode into a tube that was connected to a sealed gas bottle (120 mL) that was periodically analyzed for hydrogen. Coulombic efficiencies using this system were similar to those obtained with the two bottle system, but the overall hydrogen recovery was reduced to 60–73%, reducing the effectiveness of this system. The low recovery of hydrogen was thought to be due to loss of H_2 through the tubing connecting the cathode chamber and the collection bottle, and diffusion through the membrane which was much larger in comparison to the reactor size here than in the two-bottle system.

A cylindrical two-chamber MEC system was tested by Rozendal *et al.* (2006a) consisting of two large disc-shaped anode and cathode chambers separated by a CEM, with each chamber 29 cm in diameter and 5 cm long (total volume of 3.3 L) (Fig. 8.3). The anode was graphite felt while the cathode was a Ti mesh coated with 50 g/m^2 of Pt. With an applied voltage of 0.5 V, they achieved a high Coulombic efficiency of $C_E = 92 \pm 6.3\%$, but the recovery of hydrogen at the cathode from the measured current was only $r_{Cat} = 57 \pm 0.1\%$, resulting in an overall hydrogen recovery of $r_{H2} = 53 \pm 3.5\%$ (2.11 ± 0.14 mol-H_2/mol-acetate). It is likely the low hydrogen recovery was due to hydrogen losses through the relatively large CEM over the duration of the experiment. In a later study this group moved the cathode onto the CEM, forming a gas diffusion electrode

(*Rozendal et al.* 2007). They referred to this as a single chamber reactor, although a second chamber is still needed for gas collection. Overall H_2 recovery was low with either a CEM or an anion exchange membrane ($r_{H2} = 23\%$).

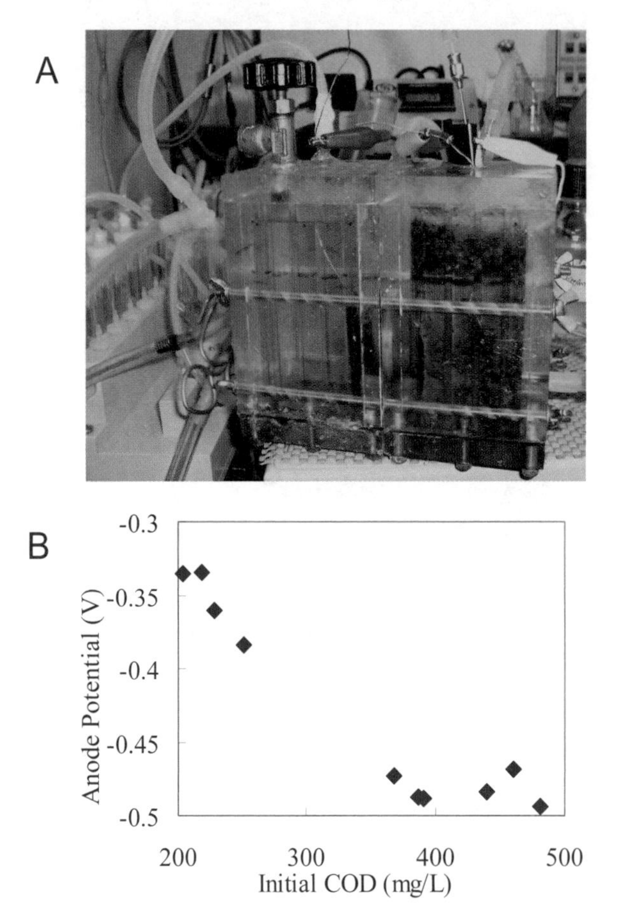

Fig. 8.4 (A) BEAMR reactor examined by Ditzig *et al.* (2007) for hydrogen production from wastewater. The anode chamber contains graphite granules and the cathode chamber has a valve to release hydrogen gas. (B) Anode potential (vs. Ag Ag/Cl electrode) decreased in proportion to wastewater strength, and therefore the performance of the reactor was limited by the relatively low COD concentration of the wastewater. (Reprinted with permission, from the International Journal of Hydrogen Energy.)

The production of H_2 from organic matter in domestic wastewater using an MEC was examined by Ditzig *et al.* (2007) using a two-cube chamber with the anode chamber containing a carbon paper electrode filled with conductive graphite granules 2–6 mm in diameter (Fig. 8.4). The CEM had a projected surface area of 11.4 cm^2, and it separated the 292-mL chambers. The minimum voltage needed to produce hydrogen was 0.23 V, but larger voltages needed to be used to obtain measurable amounts of H_2. The efficiency of the reactor as a method of wastewater treatment was excellent, with overall BOD removals of 97 ± 2% (n = 8; range 90% to 100%) and COD removals of 95 ± 2% (n = 10; range 89% to 97%). However, the performance of the reactor as a method of hydrogen production was poor. Hydrogen Coulombic efficiencies using the wastewater ranged from C_E = 9.6% to 26.2%. Electrons transferred to the cathode chamber and recovered as H_2 ranged from r_{Cat} = 1.8% to 42.7%. At an applied voltage of 0.5 V, the cathodic hydrogen recovery was 42.7% and the C_E was 23%, resulting in an overall hydrogen recovery of r_{H2} = 9.8%, or a hydrogen yield of Y_{H2} = 0.0125 mg H_2/mg-COD.

The low C_E using wastewater can be expected based on MFC tests where it has been found that in general the C_E using a wastewater is lower than that obtained with single compounds like acetate. However, another problem encountered with the domestic wastewater sample was the relatively low concentration of organic matter in the wastewater, compared to much higher values typically used in MFC and MEC tests (1 g/L of substrate or more). As shown in Fig. 8.4B, the anode potential decreased inversely with wastewater strength, and therefore only at a high COD concentration was the anode with wastewater at potential (−0.5 V vs. Ag Ag/Cl or −0.3 V vs. NHE) comparable to that achieved with acetate (−0.3 V vs. NHE).

8.3 Hydrogen yield

The performance of the MEC reactor for hydrogen generation can be evaluated using different methods. The hydrogen yield based on COD removal, Y_{H2} (mg-H_2/mg-COD) is calculated as

$$Y_{H2} = \frac{n_{H2} M_{H2}}{v_L \, \Delta COD} \tag{8-3}$$

where n_{H2} is the moles of hydrogen recovered in the experiment, calculated from the ideal gas law based on the volume of hydrogen recovered, M_{H2} is the molecular weight of hydrogen, and v_L the volume of liquid in the anode chamber. ΔCOD_i is the change in COD based on the concentrations in the reactor influent (COD_i) and effluent (COD_e) for continuous flow tests, or the starting and final *COD*s for batch tests. If the volume of hydrogen, v_{H2}, is measured (at 1 atm), then the moles of hydrogen are calculated as

$$n_{H2} = \frac{v_{H2} \, p}{R T} \tag{8-4}$$

where P (atm) is the pressure, R the gas constant (0.08206 L-atm/mol-K), and T the absolute temperature (303 K) at the time of gas sampling.

For known substrates, it is easier to base the calculation on the mass of substrate used. Assuming a change in the mass concentration of substrate, c_S, with a molecular weight of M_S, the yield Y_{H2} (g-H_2/g-substrate) is

$$Y_{H2} = \frac{n_{H2} M_{H2}}{v_L \, \Delta c_S} \tag{8-5}$$

Table 8.1 Summary of conversions for hydrogen production

Yield	Glucose—complete	Glucose—fermentation
Molar (mol-H_2/mol-glu)	12	4
Mass (g-H_2/g-COD)	0.126	0.0419
(g-H_2/kg-COD)	126	42
Volumetric: 0°C (mL/g-COD)	1410	470
30°C (mL/g-COD)	1570	520

For glucose the maximum molar yield of H_2 is 12 mol-H_2/mol-glucose, and on the basis of COD it is $Y_{H2} = 0.126$ g-H_2/g-COD as we see in the example below.

Example 8.1

Glucose fermentation by bacteria can produce a maximum of 4 mol-H_2/mol-glu. (a) Compare this to the the maximum molar and mass yields for hydrogen production in a MEC process. (b) Convert this to volumetric yields at 30°C.

(b) To calculate the molar yield, we balance the equation for the anaerobic oxidation of glucose to CO_2:

$$C_6H_{12}O_6 + 6\,H_2O \rightarrow 6\,CO_2 + 12\,H_2$$

Thus, we can produce 12 moles of hydrogen as a result of a removal of 24 e^- per mole of glucose. The mass yield is calculated as

$$Y_{H2} = \frac{12\,\text{mol-H}_2}{\text{mol-glu}}\frac{2\,\text{g-H}_2}{\text{mol-H}_2}\frac{1\,\text{mol-glu}}{180\,\text{g-glu}}\frac{\text{g-glu}}{1.06\,\text{g COD}} = 0.126\frac{\text{g-H}_2}{\text{g-COD}}$$

(d) To calculate the volume of hydrogen, we use the ideal gas law, and have

$$Y_{H2} = \frac{0.126\,\text{g-H}_2}{\text{g-COD}}\frac{0.0821\,\text{L atm}}{\text{mol K}}\frac{303\,\text{K}}{1\,\text{atm}}\frac{1\,\text{mol-H}_2}{2\,\text{g-H}_2}\frac{10^3\,\text{mL}}{\text{L}} = 1570\frac{\text{mL-H}_2}{\text{g-COD}}$$

These results are summarized in Table 8.1, along with the volumes at 0°C, for complete glucose oxidation versus that possible from fermentation.

8.4 Hydrogen recovery

Maximum moles of hydrogen produced. While the hydrogen yield represents the overall accomplishment of hydrogen production on the basis of COD removal, a more detailed analysis of hydrogen recovery is needed to understand system performance. For a specific substrate, the number of moles of hydrogen produced, n_{th}, is

$$n_{th} = \frac{b_{H2/s}\,v_L\,\Delta c_S}{M_S} \tag{8-6}$$

where $b_{H2/s}$ is the stoichiometric number of moles of hydrogen produced per mole of substrate, and M_S the molecular weight of the substrate. The theoretical limit in terms of the maximum (stoichiometric) yield in units of g-H_2/g-COD, based on COD removal, is

$$n_{th} = Y_{th}\, v_L\, \Delta COD \tag{8-7}$$

We can now explore where hydrogen losses occur by calculating the recoveries based on Coulomic efficiency and moles of hydrogen recovered in the cathode chamber.

Coulombic hydrogen recovery (Coulombic Efficiency). The moles of hydrogen that can recovered based on the measured current, $n_{CE,}$ is calculated as

$$n_{CE} = \frac{\int_{t=0}^{t} I\, dt}{2F} \tag{8-8}$$

where I the current calculated as previously described, 2 is the moles of electrons per mole of hydrogen, and F is Faraday's constant (96,485 C/mol-e^-). The Coulombic hydrogen recovery, r_{CE}, is calculated as

$$r_{CE} = \frac{n_{CE}}{n_{th}} = C_E \tag{8-9}$$

We see from eq. 8-9 that n_{CE} is the total moles of electrons recovered in the circuit divided by 2. Also, n_{th} is the total moles of electrons that could be produced by complete oxidation of the substrate divided by 2. The hydrogen Coulombic recovery, r_{CE}, is therefore equal to the Coulombic efficiency as the value of 2 cancels out.

Cathodic hydrogen recovery. Now that we know the number of moles of hydrogen that we should recover based on the measured current, we can examine how much of the hydrogen we actually recovered from this current as

$$r_{Cat} = \frac{n_{H2}}{n_{CE}} \tag{8-10}$$

where r_{Cat} is the cathodic hydrogen recovery. In laboratory tests, Liu *et al.* (2005c) obtained values of r_{Cat} for a two-bottle reactor in the range of 0.90 to 1.00, Rozendal *et al.* (2006a) obtained r_{Cat}= 0.57, with values obtained in other studies reaching 100% (Table 8.2).

Overall hydrogen recovery. The efficiency of hydrogen production, on the basis of the total recovered moles of hydrogen versus that theoretically possible, is

$$r_{H2} = r_{Cat}\, C_E = \frac{n_{H2}}{n_{th}} \tag{8-11}$$

The overall hydrogen recovery has a maximum value of r_{H2} = 1 mol/mol, while the hydrogen yield for glucose has a maximum value of Y_{H2} = 0.126 g-H_2/g-COD.

Table 8.2 Hydrogen recoveries obtained in BEAMR studies

Study	Substrate	Hydrogen recoveries		
		Coulombic C_E (%)	Cathode r_{Cat} (%)	Overall r_{H2} (%)
Liu *et al.* (2005c)—0.25 V applied, bottle	Acetate	60	90	54
Liu *et al.* (2005c) —0.45 V applied, cube	Acetate	65	94	61
Rozendal *et al.* (2006a) —Discs	Acetate	92	57	53
Rozendal *et al.* (2007) —SC	Acetate	23	101	23
Ditzig *et al.* (2007) —cube	Wastewater	9.6–26	1.8–43	9.8
Cheng and Logan—0.6 V applied (unpublished)	Acetate	87.5	100	87.5

[a]Cannot be calculated due to unknown substrate composition of wastewater.

8.5 Energy recovery

Energy recovery based on electricity input. The energy input, W_{in} (kWh), to be added into the system is determined by integrating the product of the voltage added at each measured current over the duration of the experiment, as

$$W_{ps} = \int_{t=0}^{t} I\, E_{ps} dt \qquad (8\text{-}12)$$

where $I = E/R_{ex}$ is the current calculated in the MFC circuit based on measuring the voltage across an external resistor (R_{ex}), E_{ps} the applied voltage of the power source, and dt the time increment (with the integration usually performed over n data points measured for the experiment over time intervals Δt). If the current is constant over time t, then this is simply $W_{ps} = I\, E_{ps}\, \Delta t$. Using an external resistor (R_{ex}) in the circuit results in a loss of power due to this method of measuring current, but this power does not go into H_2 production. With some power sources, this loss of power can be automatically corrected (*Rozendal et al.* 2007). For other power sources, we must adjust the calculated power requirement by subtracting energy lost by the inclusion of the resistor as

$$W_{in} = W_{ps} - W_R = \int_{t=0}^{t} \left(I\, E_{ps} - I^2 R_{ex} \right) dt \qquad (8\text{-}13)$$

The loss of power by the external resistor should be minimized by choosing a low-ohm resistor, resulting in loss of energy of only a few percent of the input energy.

The energy input for the process can be converted to the equivalent number of moles of hydrogen on an energy basis, n_{in}, using the energy content of hydrogen calculated from its heat of combustion, ΔH_{H2} (kJ/mol), as

$$n_{in} = \frac{W_{in}}{\Delta H_{H2}} \qquad (8\text{-}14)$$

where a unit conversion of 3600 kJ/kWh is needed here for the given units of ΔH_{H2} (kJ) and W_{in} (kWh). For hydrogen, we use ΔH_{H2} = 285.83 kJ/mol (the enthalpy, or upper heating value), to be consistent with manner in which values will be defined for the substrate. Note that this upper heating value for hydrogen gas assumes water in the liquid phase, while the lower heating value of 242 kJ/mol = 121 kJ/g is based on water in the gas phase.

Table 8.3 Energy recoveries and reactor production evaluated in terms of current and hydrogen normalized by reactor volume, on the basis of the upper heating value of hydrogen, for studies with acetate as the substrate

		Energy recoveries		Production	
Study	Energy input of electricity e_{in} (%)	Electricity η_W (%)	Electricity + substrate η_{W+S} (%)	I_V (A/m^3)	Q_{H2}[a] (m^3/m^3-d)
Liu *et al.* (2005c) — 0.25 V applied, bottle	12	533	62	0.45	0.0045
Liu *et al.* (2005c) — 0.45 V applied, cube	21	309	64	35	0.37
Rozendal *et al.* (2006a) —Discs	29	169	49	2.8	0.02
Rozendal *et al.* (2007) —SC	17	148	25	28[b]	0.30[b]
Cheng and Logan —0.6 V (unpublished)	30	263	80	99	1.10

[a]Calculated at 303 K.
[b]Based only on anode chamber; cathode chamber is not an integral part of the system.

The efficiency of the process relative to only the electricity input, η_W, is calculated as the ratio of the hydrogen produced to the energy content of the hydrogen recovered, or

$$\eta_W = \frac{n_{H2}}{n_{in}} \tag{8-15}$$

The value of $1/\eta_W$ is the fraction of the produced hydrogen that was due to the energy in the electricity. Note that we could also express this efficiency based on the energy required and the energy content of the hydrogen produced as

$$\eta_W = \frac{W_{H2}}{W_{in}} = \frac{n_{H2}\,\Delta H_{H2}}{W_{in}} \tag{8-16}$$

With a few assumptions, the above equation can be simplified and the related directly to the applied voltage and the hydrogen recovered. Using the moles of hydrogen recovered as defined in eq. 8-8, and assuming the current is constant, $n_{H2} = n_{CE} r_{Cat} = I \Delta t r_{Cat} / 2F$. For this conditions the power input is also assumed constant, so $W_{ps} = IE_{ps} \Delta t$ for the case of no resistor in the circuit. Using these results in eq. 8-16, we have

$$\eta_W = \frac{\Delta H_{H2} r_{Cat}}{2F E_{ps}} = \frac{\Delta H_{H2}(kJ/mol) r_{Cat}}{193 E_{ps}(V)} \tag{8-17}$$

In tests using acetate by Liu *et al.* (2005c), it was reported that there was an average energy equivalent to 0.20 mol of hydrogen was needed per mol of hydrogen produced, resulting in η_W = 500%. However, this result was based on average recoveries (not a specific result), and the lower heating value of hydrogen was used for this calculation. The energy recovery varies with the applied voltage (see example below). For the lowest voltage of 0.25 V, Liu *et al.* (2005c), obtained a hydrogen recovery of r_{Cat}= 90%, and thus an energy recovery of η_W = 533% (Table 8.3). In tests using domestic wastewater, hydrogen yields were low relative to energy input, with energy recoveries ranging from only η_W= 108% to 93% (*Ditzig et al.* 2007). With values of η_W< 100%, there is more energy input using the power source than is than that recovered as a product gas.

This energy efficiency for the MEC process based on electricity input can be compared to typical values for water electrolysis, where the efficiency is always less than unity. The energy equivalent to 1.5 to 1.7 moles of hydrogen must be input per mole of hydrogen produced (*Liu et al.* 2005c), so the energy efficiency on this basis is η_W = 59–67%, which is consistent with a generally reported range of 50–70%. Commercial water electrolyzers are often optimized for total cost and production rate, not just energy efficiency. Thus, they often operate at lower energy efficiencies in order to minimize costs for materials or to operate at specific design rates. Energy efficiencies greater than unity for the MEC are made possible by the energy contained within the organic matter, and the fact that organic matter is "electrolyzed" by bacteria and not water. The input energy really just serves to recombine the protons and electrons to form H_2 gas, a process known in the literature as the hydrogen evolution reaction (HER).

Example 8.2

Using eq. 8-17, calculate (a) how the energy efficiency input varies with the applied voltage for the case where 100% of the hydrogen produced in the cathode chamber is recovered, and (b) the point where the energy input is equal to that of the hydrogen produced.

(a) In tests by Liu *et al.* (2005c), a minimum of 0.13 V was indicated to be needed for hydrogen evolution (see Fig. 8.1A). Starting with this value, and using eq. 8-17, we can plot the efficiency as a function of applied voltage as shown below.

(b) When η_W = 100%, the energy in put is equal to the energy content of the hydrogen gas, so solving for the value of the applied potential we have

$$E_{ps}(\mathrm{V}) = \frac{\Delta H_{\mathrm{H2}}(\mathrm{kJ/mol})\, r_{Cat}}{193\,\eta_W} = \frac{(285.83\,\mathrm{kJ/mol})(1)}{193(1)} = 1.48\ \mathrm{V}$$

Thus, above an applied voltage of 1.48 V there will be less energy recovered in the hydrogen gas relative to that input as electricity. At an applied voltage of 1.8 V, or a voltage comparable to that used for water electrolysis, the efficiency drops to $\eta_W = 82\%$. Thus, there is little reason to use a MEC process when it is necessary to apply much more than one volt.

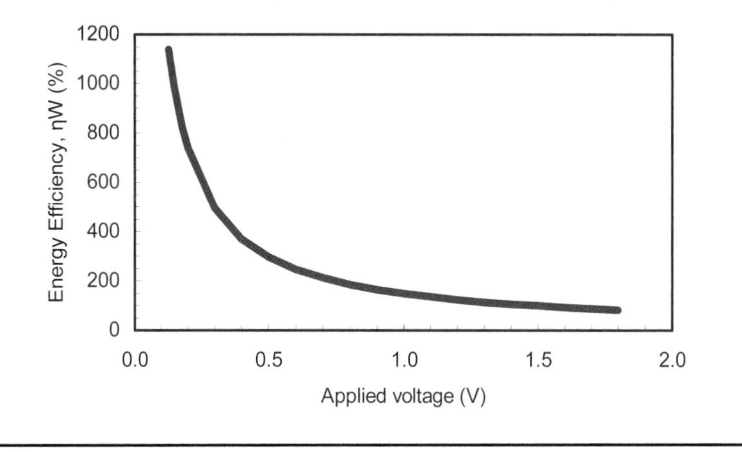

Energy recovery based on substrate and electricity input. The recovery of energy calculated above does not consider the energy input contributed by the substrate degraded by the bacteria. We can calculate an overall energy recovery of the MEC process, η_{W+S}, based on both the substrate energy content and input voltage as

$$\eta_{W+S} = \frac{W_{\mathrm{H2}}}{W_{in} + W_s} \tag{8-18}$$

where W_{H2} is the energy value of the recovered as hydrogen, calculated as

$$W_{\mathrm{H2}} = n_{\mathrm{H2}}\, \Delta H_{\mathrm{H2}} \tag{8-19}$$

Note that the units for W_{req} are in kWh, so a conversion factor of 3600 kJ/kWh is needed. W_S is the energy content of the substrate, which is similarly calculated as

$$W_S = \Delta H_S\, n_S \tag{8-20}$$

where ΔH_S is the heat of combustion for the substrate (kJ/mol), and n_S is the moles of substrate. For acetate, we can calculate ΔH_S based on the reaction

$$\mathrm{CH_3COOH} + 2\,\mathrm{O_2} \rightarrow 2\,\mathrm{CO_2} + 2\,\mathrm{H_2O} \tag{8-21}$$

Using tabulated values of −393.51 kJ/mol for CO_2, −285.83 kJ/mol for H_2O, −488.40 kJ/mol for acetate, and 0 kJ/mol for O_2, we calculate the heat of combustion as

$$\Delta H_S = [2(-393.51) + 2(-285.83)] - [1(-488.40)] = -870.28 \frac{\text{kJ}}{\text{mol}} \tag{8-22}$$

An alternative method for calculating η_{W+S} is based on substituting hydrogen recovery and energy efficiency values into eq. 8-18. The energy content of the hydrogen produced (W_{H2}) is given by eq. 8-19, and the electrical energy (W_{in}) input is given by eq. 8-13. The energy content of the substrate is shown above in eq. 8-20, but we recognize that the moles of substrate used is related to the moles of hydrogen produced by $n_S = n_{H2}/(b_{H2/s}\, r_{H2})$, where n_{H2}/r_{H2} is the number of moles of hydrogen that could have been made from the substrate, and $b_{H2/s}$ is the stoichiometric moles of hydrogen possible from the number of moles of substrate. Thus, we see that

$$W_S = \frac{\Delta H_S\, n_{H2}}{b_{H2/s}\, r_{H2}} \tag{8-23}$$

Combining these results (eqs. 8-16, 8-19, and 8-23), eq. 8-18 becomes

$$\eta_{W+S} = \left(\frac{1}{\eta_W} + \frac{\Delta H_S}{\Delta H_{H2}} \frac{1}{b_{H2/s}\, r_{H2}} \right)^{-1} \tag{8-24}$$

Using the above information, we can also evaluate the process solely on the efficiency of hydrogen production based on the hydrogen produced and the substrate used, as

$$\eta_S = \frac{W_{H2}}{W_S} = \frac{\Delta H_{H2} b_{H2/s}\, r_{H2}}{\Delta H_S} \tag{8-25}$$

Based on the hydrogen recoveries shown in Table 8.3, we have energy recoveries as high as 80% in this process. Note that values for Liu *et al.* were re-calculated based on values given in their paper, and using ΔH_{H2} = 285.83 kJ/mol as done for the other calculated values.

Another way to examine the efficiency of the process is to compare energy input in the form of electricity or substrate as a proportion of the total energy input. The energy required for electricity relative to that for both electricity and substrate, is

$$e_{in} = \frac{W_{in}}{W_{in} + W_S} \tag{8-26}$$

Similarly, the energy input for the substrate is

$$e_S = \frac{W_S}{W_{in} + W_S} = 1 - e_{in} \tag{8-27}$$

Reactor performance based on volumetric densities. The final metric for achieving optimal reactor performance is the hydrogen production rate, evaluated in terms of current produced per volume of reactor and the gas rate per volume. Reported current densities in units of A/m^2-electrode can be converted to volumetric densities, I_v, (A/m^3) using the electrode sizes and reactor volume. From the volumetric current density, with 2 mol-e^- produce 1 mol H_2, Faradays constant and the ideal gas law, the volumetric H_2 production rate, Q_{H2} (m^3/m^3-d), is calculated using the dimensionless cathodic hydrogen recovery with the indicated units as

$$Q_{\mathrm{H2}}(\frac{\mathrm{m}^3}{\mathrm{m}^3\,\mathrm{d}}) = \frac{I_V\,(\mathrm{A/m^3})\,r_{Cat}[(1\mathrm{C/s})/\mathrm{A}]\,(0.5\ \mathrm{mol\,H_2/mol\,e^-})\,(86{,}400\,\mathrm{s/d})}{(F = 9.65\times10^4\,\mathrm{C/mol\,e^-})\,c_g\,(\mathrm{mol\,H_2/L})\,(10^3\,\mathrm{L/m^3})} = \frac{43.2 I_V\, r_{Cat}}{F\,c_g(\mathrm{T})} \tag{8-28}$$

where c_g is the concentration of gas at a temperature T calculated using the ideal gas law (0.040 mol/L at 30°C), and 43.2 is for unit conversions. We can equivalently express this relationship by substituting in the ideal gas law ($c_g = p/RT$), assuming hydrogen is produced at 1 atm of pressure, and by including F into the constant, to produce

$$Q_{H2}(\frac{\mathrm{m}^3}{\mathrm{m}^3\,\mathrm{d}}) = 3.68\times10^{-5} I_V\,(\frac{\mathrm{A}}{\mathrm{m}^3})\,T(\mathrm{K})\,r_{Cat} \tag{8-29}$$

The accomplishments made to date in terms of energy recovery and the rate of hydrogen production are shown in Fig. 8.5. Using a two-chamber bottle reactor, Liu *et al.* (2005c) produced a volumetric current density of 0.45 A/m^3, or a production rate of 0.0045 m^3-H_2/d-m^3-based on total reactor volume (anode and cathode). These rates were

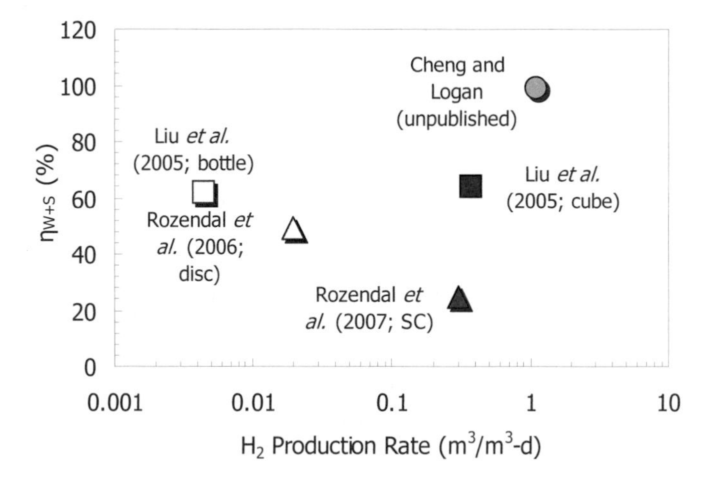

Fig. 8.5 Energy efficiencies (based on electricity and substrate input) reported in several studies compared to the volumetric hydrogen production rate in MEC experiments.

increased by using a cube-reactor design to 35 A/m^3, but overall hydrogen recover was only 46% (Table 8.2). Other designs examined by Rozendal (*Rozendal et al.* 2007; *Rozendal et al.* 2006a) were within the range of values obtained by Liu *et al.* (2005c) in terms of overall energy efficiency and current density. In recent tests, Cheng and Logan (unpublished) achieved a hydrogen production rates of 1.1 m^3/d-m^3 (99 A/m^3) using a new type of reactor design (Fig. 8.5 and Table 8.3). Note that these rates are normalized to the volume of the whole reactor (both the anode and cathode chamber volumes). In many cases, however, the cathode volumes were not optimized but rather set equal to the anode chamber size. Thus, these rates could likely double with more optimized cathode designs relative to volumes.

These hydrogen production rates can be compared to those achieved in fermentation systems on the same basis of volumetric hydrogen production per day, normalized to reactor volume. Hawkes *et al.* (2007) identified 12 studies with sucrose or glucose as the substrate where hydrogen production rates varied over a wide range of 0.15 to 15.1 m^3/m^3-d (average of 3.1 ± 4.6 m^3/m^3-d). Except for one high rate of 15.1 m^3/m^3-d, production rates for 11 studies averaged 2.0 ± 2.8 m^3/m^3-d. Lower rates were reported using wastewaters containing high concentrations of sugar. Hydrogen production rates with MEC have exceeded these lower rates, and are rapidly approaching these upper rates. Thus, it is easy to imagine coupling fermentation and MEC reactors for substrates such as easily fermented sugars. However, the MEC production of hydrogen is not limited to sugar substrates as any biodegradable material can be used in the process. Thus, we see for the first time that we can achieve hydrogen production rates that can be useful for generation of hydrogen from renewable biomass resources. With the improvements being made in MFC systems, it should be possible to continue to improve MEC designs and increase these production rates.

Example 8.3

MEC experiments were conducted with acetate that provided the following information: moles of hydrogen produced, $n_{H2} = 0.34$ mmol (3.8 mol-H_2/mol-Ac); Coulombic hydrogen recovery, $C_E = 0.875$; cathodic hydrogen recovery, $r_{Cat} = 1.0$; energy required, $W_{req} = 0.037$ kJ; current density, $I_v = 99$ A/m^3. (a) Calculate the efficiencies of energy recovery based on electricity, substrate, and electricity for acetate (ΔH_S = 870.28 kJ/mol). (b) Calculate the volumetric hydrogen production rate in m^3/d-m^3.

(a) First we need to calculate the overall hydrogen recovery from eq. 8-11 as

$$r_{H2} = C_E \, r_{Cat} = (0.875)\,(1.0) = 0.875$$

Next, we can calculate the different efficiencies with the maximum molar yield of $b_{H2/s} = 4$ mol-H_2/mol and using the upper heat of combustion for H_2 of 285.83 kJ/mol. From eqs. 8-16, 8-24, and 8-25, we have:

$$\eta_W = \frac{W_{H2}}{W_{in}} = \frac{n_{H2}\,\Delta H_{H2}}{W_{in}} = \frac{(3.4\times10^{-4}\text{ mol H}_2)(285.83\text{ kJ})}{(0.037\text{ kJ})}\times 100\% = 263\%$$

$$\eta_S = \frac{W_{H2}}{W_S} = \frac{\Delta H_{H2} b_{H2/s}\, R_{H2}}{\Delta H_S} = \frac{(285.83\text{ kJ})\,(4\text{ mol/mol})\,(0.875)}{(870.28\text{ kJ})}\times 100\% = 115\%$$

$$\eta_{W+S}=\left(\frac{1}{\eta_W}+\frac{\Delta H_S}{\Delta H_{\mathrm{H2}}}\frac{1}{b_{\mathrm{H2}/s}\,r_{\mathrm{H2}}}\right)^{-1}=\left(\frac{1}{(2.63)}+\frac{(870.28\,\mathrm{kJ})}{(285.83\,\mathrm{kJ})}\frac{1}{(4\,\mathrm{mol/mol})(0.875)}\right)^{-1}\times100\%=80\%$$

From this analysis, we see that hydrogen production evaluated on the basis of either the electricity or substrate alone produces an overall efficiency of greater than 100%, but on the basis of combustion energy for the substrate and hydrogen the overall energy efficiency is 80%.

(b) The hydrogen production rate at 30°C is obtained from eq. 8-29 as

$$Q_{\mathrm{H2}}(\frac{\mathrm{m}^3}{\mathrm{m}^3\,\mathrm{d}})=3.68\times10^{-5}I_V\,(\mathrm{A/m^3})\,T\,(\mathrm{K})\,r_{\mathrm{H2}}=3.68\times10^{-5}\,(99\,\mathrm{A/m^3})\,(303\,\mathrm{K})\,(1)$$

$$Q_{\mathrm{H2}}=1.10\,\frac{\mathrm{m}^3}{\mathrm{m}^3\,\mathrm{d}}$$

Hydrogen production is thus quite favorable in terms of rates typical of fermentation processes.

Including the energy efficiency of the power source. In the above energy efficiency calculations, we include energy input for the power source, but that voltage is generated at some energy efficiency as well. To include this energy efficiency, we can divide the energy required by the power source by the power source energy efficiency, η_{ps}. For example, including this term in the efficiency based on electricity input, eq. 8-16 becomes

$$\eta_W=\frac{W_{\mathrm{H2}}}{W_{in}/\eta_{ps}}=\frac{n_{\mathrm{H2}}\,\Delta H_{\mathrm{H2}}}{W_{in}/\eta_{ps}} \tag{8-30}$$

If electricity is produced by combustion, we might have an efficiency of 33% for that process. In Example 8.3, this would reduce the overall energy efficiency from η_W= 263% to just η_W= 87%. However, if hydrogen evolved from the process was used in a hydrogen fuel cell operating at 50% efficiency, then this would become η_W= 132%. If we use the same approach based on energy efficiency from electricity and substrate, eq. 8-24 becomes

$$\eta_{W+S}=\left(\frac{1}{\eta_W\eta_{ps}}+\frac{\Delta H_S}{\Delta H_{\mathrm{H2}}}\frac{1}{b_{\mathrm{H2}/s}\,r_{\mathrm{H2}}}\right)^{-1} \tag{8-31}$$

For our case above in Example 8.3, an efficiency of the power source of 50% for electricity production would result in an overall efficiency of 61%. This is still quite good compared to the case for water electrolysis, which would become 17–23% for the case using energy generated by a combustion process, and 25–35% for a hydrogen fuel cell. Of course using other renewable forms of energy production, including solar, wind or MFCs, would result in different outcomes.

8.6 Hydrogen losses

The loss of hydrogen from the cathode chamber in the MEC process occurs from several different processes, including: (a) diffusion to the anode chamber through the water and membrane separating the chambers; (b) microbiological uptake of hydrogen due to methanogenesis and use of alternate electron acceptors such as sulfate and nitrate; (c) abiotic conversion of hydrogen to methane (it is thermodynamically favorable). Here we explore the potential for mass transfer losses due to hydrogen diffusion through a CEM, using Nafion as an example.

Mass transfer coefficients. The overall flux of hydrogen from the cathode chamber to the anode chamber in the MEC reactor is

$$J_{CA} = \boldsymbol{K}_{CA}\left(c_{\mathrm{H2},Cat} - c_{\mathrm{H2},An}\right) \tag{8-32}$$

where J_{CA} is the flux (mol/cm^2-s), $\boldsymbol{K}_{Hca}$ the overall mass transport coefficient (*Logan* 1999) for hydrogen transport between the cathode and anode, and $c_{\mathrm{H2},Cat}$ and $c_{\mathrm{H2},An}$ are the concentrations of hydrogen in water in the cathode and anode chambers, respectively. We assume that due to biological consumption of H_2 in the anode chamber that $c_{An} = 0$ (this also gives the maximum possible rate of hydrogen loss possible by mass transfer). The idea here is that the flux of hydrogen from the system is constant (just like we assume oxygen intrusion into the anode chamber is constant in an MFC calculation of Coulombic efficiency losses). Thus, a goal of MEC reactor design is to minimize hydrogen losses by keeping hydrogen generation distant from the membrane.

The overall mass transfer coefficient can be separated into two resistances,

$$\frac{1}{\boldsymbol{K}_{CA}} = \frac{1}{\boldsymbol{k}_{\mathrm{H2}w}} + \frac{1}{\boldsymbol{k}_{\mathrm{H2}m}} \tag{8-33}$$

where $\boldsymbol{k}_{\mathrm{H2}w}$ and $\boldsymbol{k}_{\mathrm{H2}m}$ are the individual mass transport coefficients for hydrogen transport across the bulk water phase. Assuming a stagnant film (*Logan* 1999), the mass transfer coefficients can be related to the chemical diffusivities as

$$\boldsymbol{k}_{\mathrm{H2}w} = \frac{D_{\mathrm{H2}w}}{\delta_w} \tag{8-34}$$

$$\boldsymbol{k}_{\mathrm{H2}m} = \frac{D_{\mathrm{H2}m}}{\delta_m} \tag{8-35}$$

where $D_{\mathrm{H2}w}$ and $D_{\mathrm{H2}m}$ are the diffusion coefficients of hydrogen in the water phase and hydrogen in the membrane phase, respectively, with hydrogen transfer occurring through the bulk phase a distance δ_w (water phase) and for hydrogen transport through the membrane of thickness δ_m (membrane phase).

Concentration of hydrogen at the cathode ($c_{\mathrm{H2}c}$). If we assume pure hydrogen is produced at the cathode, then the concentration of hydrogen in water at the cathode surface would be that concentration in equilibrium with pure hydrogen gas. From the ideal gas law at 1 atmosphere, we know that the hydrogen gas concentration is $n/V = p/RT = c_{\mathrm{H2}g} = 0.0402$ mol/L (303 K). Using Henry's law, where is the Henry's law constant of hydrogen is H_{H2} = 52.76 (mol/L-gas)/(mol/L-water) (303 K), we calculate the concentration of hydrogen in water (*Lide* 1995) at the cathode as

$$c_{H2,Cat} = \frac{c_{H2g}}{H_{H2}} = \frac{(0.0402\ \text{mol/L - gas})}{[52.76\ (\text{mol/L - gas})/(\text{mol/L - water})]}$$

$$= 7.62 \times 10^{-4} \frac{\text{mol}}{\text{L - water}} \tag{8-36}$$

Diffusion coefficient of hydrogen in water. At 298 K, the diffusion constant of hydrogen in water is 5.85 × 10^{-5} cm^2/s (*Perry and Chilton* 1973). Correcting for temperature using $D_{H2w}\mu_w/T$ = constant (*Logan* 1999) where μ_w is the dynamic viscosity of water at temperature T (K), we calculate at 303 K the diffusion coefficient is

$$D_{H2w}\ (303\ \text{K}) = 6.64 \times 10^{-5}\ \text{cm}^2/\text{s} \tag{8-37}$$

Diffusivity of hydrogen in the membrane. The diffusion coefficient of hydrogen is calculated for the membrane, based on data for water saturated Nafion, at 293 K and 313 K of 7.6 × 10^{-6} cm^2/s and 1.29 × 10^{-5} cm^2/s (*Jiang and Kucernak* 2004), respectively. The diffusion coefficient at 303 K by linear interpolation is

$$D_{H2m}(303\ \text{K}) = 1.03 \times 10^{-5} \frac{\text{cm}^2}{\text{s}} \tag{8-38}$$

Example 8.4

The reactor shown below was used in tests for hydrogen production using domestic wastewater (*Ditzig et al.* 2007). Based on this system set up, calculate (a) the maximum hydrogen flux of hydrogen through the membrane assuming a membrane thickness of 0.0183 cm for Nafion 117 and a distance between the membrane and electrode of 8.6 cm, and (b) the volume of hydrogen lost over a 40-h batch cycle for a membrane surface area of A_m = 11.9 cm^2. Compare this volume to that found (9.2 mL). (c) Repeat the calculation with the cathode pressed next to the membrane.

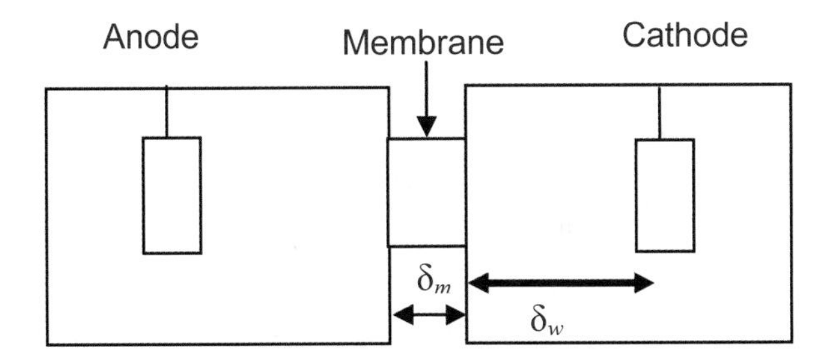

(a) The water phase mass transfer coefficient is calculated using δ_w = 8.60 cm as the distance between the cathode and membrane in the MEC reactor, is

$$\boldsymbol{k}_{\mathrm{H2}w} = \frac{D_{\mathrm{H2}w}}{\delta_w} = \frac{6.64\times10^{-5}\ \mathrm{cm^2/s}}{8.60\ \mathrm{cm}} = 7.72\times10^{-6}\ \frac{\mathrm{cm}}{s}$$

For the membrane, we similarly calculate the membrane mass transfer coefficient $\boldsymbol{k}_{H2m}$, as

$$\boldsymbol{k}_{\mathrm{H2}m} = \frac{D_{\mathrm{H2}m}}{\delta_m} = \frac{1.03\times10^{-5}\mathrm{cm^2/s}}{0.0183\ \mathrm{cm}} = 5.63\times10^{-4}\ \frac{\mathrm{cm}}{\mathrm{s}}$$

The overall mass transfer coefficient, $\boldsymbol{K}_{CA}$, is calculated by substitution of $\boldsymbol{k}_{\mathrm{H2}w}$ and $\boldsymbol{k}_{\mathrm{H2}m}$ into eq. 8.33, which produces here

$$\frac{1}{\boldsymbol{K}_{CA}} = \frac{1}{\boldsymbol{k}_{\mathrm{H2}w}} + \frac{1}{\boldsymbol{k}_{\mathrm{H2}m}} = \frac{1}{7.72\times10^{-6}\ \mathrm{cm/s}} + \frac{1}{5.63\times10^{-4}\ \mathrm{cm/s}}$$

$$\boldsymbol{K}_{CA} = 7.62\times10^{-6}\ cm/s$$

Thus, we see that most of the resistance (99%) to mass transfer is due to the bulk phase resistance. The flux of hydrogen from the cathode to the anode can now be found, assuming that $c_{\mathrm{H2},An}$ is equal to zero. Although wastewater is not devoid of dissolved hydrogen, it is assumed that the dissolved hydrogen is zero to find a maximum flux:

$$J_{CA} = \boldsymbol{K}_{CA}\, c_{\mathrm{H2},Cat} = (7.62\times10^{-6}\ \mathrm{cm/s})\,(7.62\times10^{-4}\mathrm{mol/L})\frac{1\ \mathrm{L}}{1000\ \mathrm{mL}}$$

$$J_{CA} = 5.80\times10^{-12}\ \frac{\mathrm{mol}}{\mathrm{cm^2\ s}}$$

(b) For a batch test of t_b = 40 h and using the given surface area of the membrane of A_m = 11.9 cm^2, the total volume of hydrogen lost over the batch cycle, v_{H2}, is

$$v_{\mathrm{H2}} = \frac{J_{CA}\,A_m\,t_b}{c_{\mathrm{H2}g}} = \frac{(5.80\times10^{-12}\mathrm{mol/cm^2\ s})\,(11.9\mathrm{cm^2})\,(40\ \mathrm{h})}{(0.0402\ \mathrm{mol/L})}\,\frac{3600\ \mathrm{s}}{\mathrm{h}}\,\frac{10^3\ \mathrm{mL}}{1\ \mathrm{L}} = 0.25\ \mathrm{mL}$$

Comparing this volume to the 9.2 mL of hydrogen recovered in the cathode chamber (under conditions where C_E = 37.5%), and conclude that only 2.7% (0.21 mL of 9.41 produced) of the hydrogen was lost due to mass transfer processes, which is quite small (*i.e.*, r_{Cat} = 97.3%). This is an upper limit as this assumes that hydrogen is always present at its solubility limit over the complete cycle.

(c) If we place the cathode against the membrane, there is no water phase resistance. We can therefore consider the overall mass transfer coefficient as that just due to the membrane. For this case, we would have

$$J_{CA} = \boldsymbol{K}_{CA}\, c_{\mathrm{H2},Cat} = (5.63\times10^{-4}\ \mathrm{cm/s})\,(7.62\times10^{-4}\mathrm{mol/L})\frac{1\ \mathrm{L}}{1000\ \mathrm{mL}}$$

$$J_{CA} = 4.29\times10^{-10}\ \frac{\mathrm{mol}}{\mathrm{cm^2\ s}}$$

$$v_{\mathrm{H2}} = \frac{(4.29\times10^{-10}\ \mathrm{mol/cm^2s})\,(11.9\mathrm{cm^2})\,(40\ \mathrm{h})}{(0.0402\ \mathrm{mol/L})}\,\frac{3600\ \mathrm{s}}{\mathrm{h}}\,\frac{10^3\ \mathrm{mL}}{1\ \mathrm{L}} = 18.3\ \mathrm{mL}$$

In this case, we see that there is substantial potential for loss of hydrogen through the membrane, as 18.3 mL is greater than the hydrogen production measured for this case. Thus, keeping the cathode away from the membrane is important.

8.7 Differences between the MEC and MFC systems

The MEC and MFC systems share many similar characteristics, and therefore many findings for improving electricity generation in MFCs are expected to be applicable for increasing hydrogen production in MECs. However, there are important differences between the systems that can affect system performance relative to hydrogen recovery. First, in the MEC process loss of hydrogen due to its diffusion through the membrane (as described above) into the anode chamber or its degradation by bacteria that might grow in the cathode chamber are primary challenges. Hydrogen needs to be generated in a manner that reduces its diffusion back into the anode chamber. In the MFC process, there is no comparable process as only water is generated at the cathode. In addition, bio-utilization of the evolved hydrogen must be minimized by reducing bacterial growth in the cathode.

Second, charge is balanced in a membrane-less MFC through diffusion of protons through only water, but in a MEC process some type of membrane is needed as a barrier to hydrogen loss. In some MFCs, a cation exchange membrane (CEM) is used to exchange protons between chambers, but typically cations such as Na^+, K^+ or Ca^{++} are exchanged rather than H^+ leading to loss of protons at the cathode (*Kim et al.* 2007b; *Rozendal et al.* 2006b). In the MEC process, this becomes critical as a high concentration of protons is needed to ensure H_2 gas evolution. Using an anion exchange membrane (AEM) can allow proton conduction via negatively charged species such as phosphate anions that can be added at high concentrations. There may not be an advantage of the AEM over the CEM in cases where a gaseous diffusion membrane is used, as observed by Rozendal *et al.* (2007).

Third, substrate can reach the cathode in an MFC or cross the membrane if present. In a MEC process, this means that substrate (food) for bacteria can allow growth of bacteria in the cathode chamber. These bacteria can then use hydrogen, decreasing hydrogen yields. Minimizing substrate crossover is therefore a critical issue with a MEC processes that have two aqueous chambers. In MFC processes with air cathode (*Liu et al.* 2005b; *Liu and Logan* 2004), or a MEC reactor having a single chamber (*Rozendal et al.* 2007), there is no concern for substrate loss via diffusion to the cathode (except there still can be volatilization losses for some chemicals).

Fourth, in an MFC there is a large loss of substrate due to oxygen diffusion into the anode chamber. In the MEC process, however, this is avoided, resulting in much higher hydrogen Coulombic efficiencies (R_{CE}s) in the MEC process than C_Es in MFCs. The R_{CE} of a MEC, for example, was in the range of 60–78% in one study (*Liu et al.* 2005c), and as high as 92% in another with acetate (*Rozendal et al.* 2007) while in MFCs C_Es vary widely from 10% to 78% (depending on designs and internal resistances) using mixed cultures and acetate (*Liu et al.* 2005a; *Oh et al.* 2004).

It is clear that advances in reactor design that minimize internal resistance in MFC reactors will benefit MEC reactor designs. However, the above comparison shows that there are important differences between the MFC and MEC systems. These differences will result in design approaches that will vary for the two technologies. Moreover, the capture and purification of the evolved hydrogen gas is a complicated and energy intensive process that will affect the overall economics of the process as a sustainable and economical method of hydrogen production.

CHAPTER 9

MFCs for Wastewater Treatment

9.1 Process trains for WWTPs

Let us assume that an MFC design becomes economically and technically viable: How will the MFC be used for wastewater treatment? First, we can consider the situation of how a wastewater treatment plant (WWTP) is designed, and then see how an MFC might be used in such an application. A typical WWTP treating domestic wastewater consists of a series of unit processes all designed with specific functions to make the monitoring and treatment of the wastewater as efficient as possible. While many different arrangements of these units are possible, we will limit our discussion here to a "typical" WWTP used for domestic wastewater treatment. Process trains will be more variable for industrial wastewaters and for nutrient control. Wastewater entering a domestic WWTP is first screened to remove large debris and monitored for flow, and then the flow goes through a grit chamber with a hydraulic residence time (HRT) of 1–20 minutes (*Viessman and Hammer* 2005) where gritty particles (such as coffee grinds, bones, *etc.*) are removed in order to protect pumps (Fig. 9.1). The solids removed from the screens and grit chamber are usually collected and sent directly to a landfill.

The wastewater is then treated by physical removal of organic matter in a primary clarifier, although in small plants this unit is sometimes omitted. The purpose of the primary clarifier is to remove all particles that are settleable as this is a very cost effective method of treatment. The organic matter concentration in the wastewater is usually evaluated in terms of either its biochemical oxygen demand (BOD) in a five-day test (BOD_5) or its chemical oxygen demand (COD) in a rapid chemical oxidation test. The former reflects what can be biologically removed while the latter is a more rapid assessment of all organic matter in the water. The total BOD (or COD) can be viewed as consisting of two fractions: soluble BOD (sBOD) and particulate BOD (pBOD). Typically, 1/3 of the BOD and 1/2 the total solids entering the treatment plant are removed by primary clarification in a process that has an HRT of 1 to 3 hours (*Viessman and Hammer* 2005). Thus, the BOD_5 of the entering wastewater is reduced from ~300 mg/L to 200 mg/L. Mostly pBOD is removed in the primary clarifiers in the form of solids collected at the bottom of the tank called primary clarifier sludge. This material is sent on for further treatment in anaerobic digesters (ADs).

Next the wastewater enters a biological treatment process which typically consists of two components: a bioreactor where BOD gets converted to bacterial biomass; and a settling tank with the bacterial biomass is removed, called a secondary clarifier. This biological process is the heart of the treatment plant, and its characteristics vary widely. For our purposes here, we can consider two treatment options: activated sludge (AS); and trickling filters (TF) (Fig 9.2). The AS process consists of a large aeration tank where entering wastewater is combined with solids from the secondary clarifier. The high concentration of bacteria in the recycled solids from the clarifier (*ca.* 10,000 mg/L) results in the rapid consumption of the organic matter in the entering wastewater (~200 mg/L of BOD_5). The HRT of the aeration tank is ~4–6 hours. The wastewater then flows into the secondary clarifier where bacterial biomass settles out, and the treated wastewater that leaves must have < 30 mg/L of BOD_5 and <30 mg/L of total suspended solids (TSS). This treatment level is easily achieved with most AS sludge plants, making them highly effective at wastewater treatment. The main difficulties in operation include: setting the right recycle rates, choosing solids concentrations that make the process effective and control sludge production; and avoiding upsets of the clarification process that affect the concentration of recycled biomass. The aeration step is effective, but quite costly as aeration can account for about half of the energy consumption at a typical WWTP.

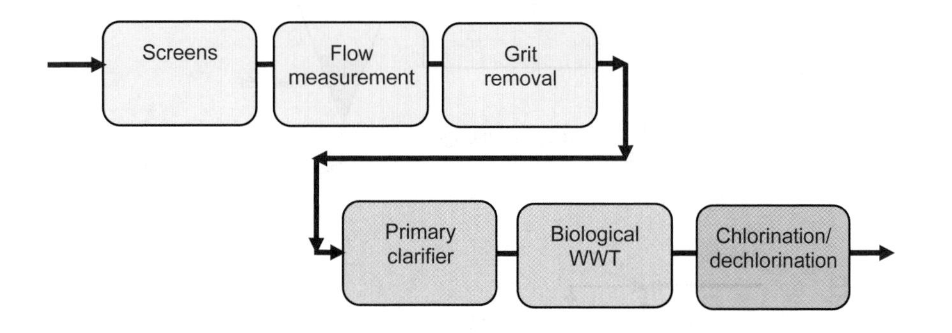

Fig. 9.1 Process flow train for a typical domestic wastewater treatment plant in the USA.

A TF process consists of large tank containing porous media where the wastewater is distributed onto the top of the medium (like a sprinkler system for a lawn). The media must maintain sufficient porosity to keep air flow through the tank so that the BOD removal is primarily aerobic BOD removal, and not anaerobic removal. TFs were first constructed using piles of slag or rocks producing porosities of 30% to 50%, but new structured plastic media used in most new systems have porosities of 90% or more. Rock filters were generally limited to heights of 2 m, but plastic media systems can range from 2 to 10 m or more in height, with tank diameters of 20 m or more. HRT in the reactor is quite difficult to determine as some wastewater flows through in 15 to 20 minutes, but dye tests indicate wastewater diffusing into the biofilm can take several hours or more. The reactors are often equipped with ventilation systems (fans) to keep air moving through the reactor. A portion of the effluent from the TF is recycled to the top of the TF to maintain constant application rates so that the biofilm remains well-wetted.

Wastewater leaving the TF enters a secondary clarifier to remove biofilm solids produced in the system. Treatment by the TF is usually not as effective as an AS process as the BOD_5 may only be <45 mg/L.

To make performance of a TF-based system comparable to that of an AS plant, a secondary solids contact (SC) process is often used, called the trickling filter/solids contact (TF/SC) process (*Parker and Bratby* 2001). In the SC process solids from the secondary clarifier underflow are recycled and combined with the TF effluent and sent to a tank with a variable but relatively short HRT (compared to AS) of an hour or less. The SC tank is aerated to promote mixing and to keep the wastewater slightly oxygenated. The wastewater then flows on to the secondary clarifier where the solids settle out, and the effluent is improved to achieve <30 mg/L BOD_5 and <30 mg/L TSS. Most of the $sBOD_5$ is removed in the TF process, with some additional $sBOD_5$ removed in the SC process. The clarifier is essential for pBOD removal. The SC process basically improves pBOD removal and thus makes the process overall more efficient. Sludge produced from the TF/SC process are usually combined with primary clarifier sludge and then treated with an anaerobic digestion process.

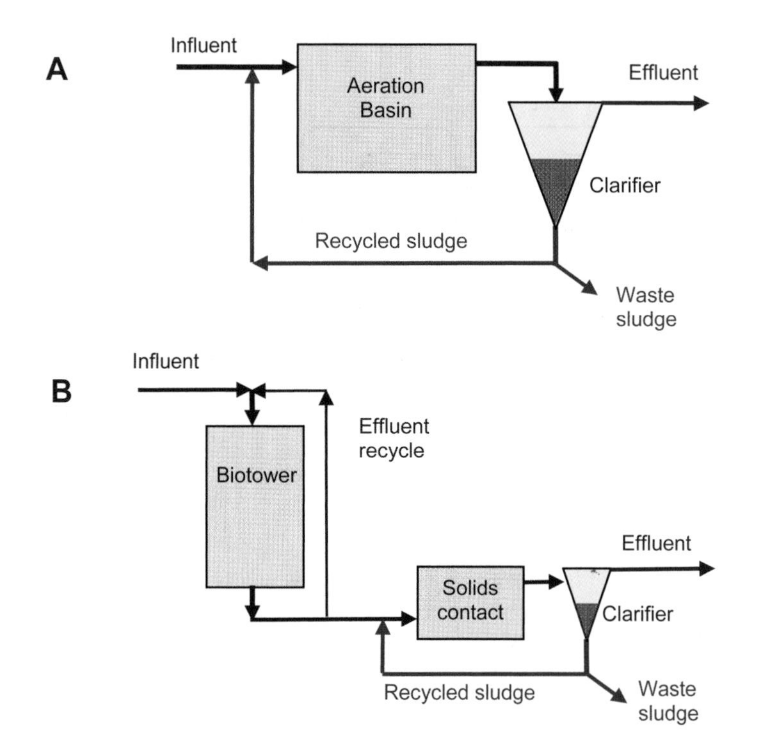

Fig. 9.2 Biological treatment process consisting of a bioreactor and secondary clarifier. (A) Activated sludge process with aeration basin, sludge recycle, and waste sludge lines; (B) Trickling filter process consisting of biotower, solids contact tank, biotower effluent recycle and sludge recycle lines.

After the wastewater leaves the biological treatment process, it is usually chlorinated in treatment plants in the US to kill bacteria, and then dechlorinated to protect the aquatic life in the receiving water body. Sometimes treated wastewater is applied to land or used for groundwater recharge, but usually it flows into a river, lake or ocean where it is further diluted. The effluent regulations for a specific treatment plant can vary based on the level of protection needed for the receiving water body. Nutrient removal (nitrogen and phosphorus) can be integrated into these biological processes or added on as separate processes, but a discussion of these treatment technologies is beyond what can be considered here in this brief summary of treatment systems. A few concluding thoughts, however, are provided at the end of this chapter.

The sludge that is produced from the clarifiers goes into a treatment train typically consisting of an anaerobic digester (AD) having a HRT of 15 to 20 days or more. The liquid effluent from the AD is recycled back to the head of the plant. The gas produced, containing primarily methane and carbon dioxide, can be used for electricity production in gas turbines with the waste heat used to heat the ADs. In some plants the gas is used to heat the digester directly with the excess methane burned (flared). The solids from this process are filter pressed to remove more water, and composted. These treated and stabilized solids, called biosolids, are now suitable for safe land application and often times can be used as a source of fertilizer.

From this summary of a WWTP process train we can see that an MFC would replace the AS or TF treatment system. The MFC is a biological treatment process, and thus its function will be to remove BOD in the same manner as accomplished by the AS aeration tank or the TF. There are four main advantages of using an MFC instead of one of these conventional bioreactors:

1. *Production of a useful product in the form of electricity.* (Or in an MEC process, hydrogen gas; see Chapter 8). The current generated is dependent on the wastewater strength and the Coulombic efficiency.

2. *Lack of a need for aeration.* Aeration in AS can consume 50% of the electricity used at a treatment plant. No aeration is needed for an air-cathode MFC that uses only passive oxygen transfer at the cathode.

3. *Reduced solids production.* The MFC is an anaerobic process, and thus bacterial biomass production will be reduced compared to that of an aerobic system such as a TF or AS. The amount of solids reduction is dependent on several assumptions as discussed below. Solids treatment is expensive, and using an MFC may substantially reduce solids production.

4. *Potential for odor control.* This is the least well studied aspect of treatment performance, but omitting high surface areas needed in TFs exposed to air, and the flow of large amounts of air through the aeration basin in the AS process could greatly reduce the potential for odor generation to a surrounding community.

9.2 Replacement of the biological treatment reactor with an MFC

It is important to keep in mind that an MFC has never been used at any scale larger than the bench-scale devices used in laboratory tests, and so at this time we have no specific data on system performance in the manner needed to specify unit processes and system conditions (HRT, surface areas, *etc.*) needed for the system to be used in an actual wastewater treatment plant. Thus, the reader should understand that this section represents the author's opinions and views—based on extrapolation of known data and his general understanding of wastewater treatment processes—about how an MFC system would be designed and implemented for domestic wastewater treatment (*Logan* 1999, 2005; *Logan et al.* 1987).

There are three possible treatment process trains envisioned. First, it is expected that the MFC process could be incorporated into the process train in a more conventional system replacing the AS or TF systems. In this case, the MFC would be used in a manner similar to that of a TF in a TF/SC arrangement (Fig. 9.3A). The MFC reactor is a biofilm process, and thus the treatment should be more effective at removing sBOD than pBOD. Thus, a SC process would be needed to remove pBOD and achieve satisfactory treatment levels for domestic wastewater treatment. Recycle of the effluent from the MFC reactor to the influent line is not expected to be needed as in a TF process as contents of the reactor would likely be completely filled (versus a TF which is open to the atmosphere). The design of the SC and clarifier processes relies on the solids from the MFC acting in the same manner as the biofilm solids from a TF, which we will assume for our calculations.

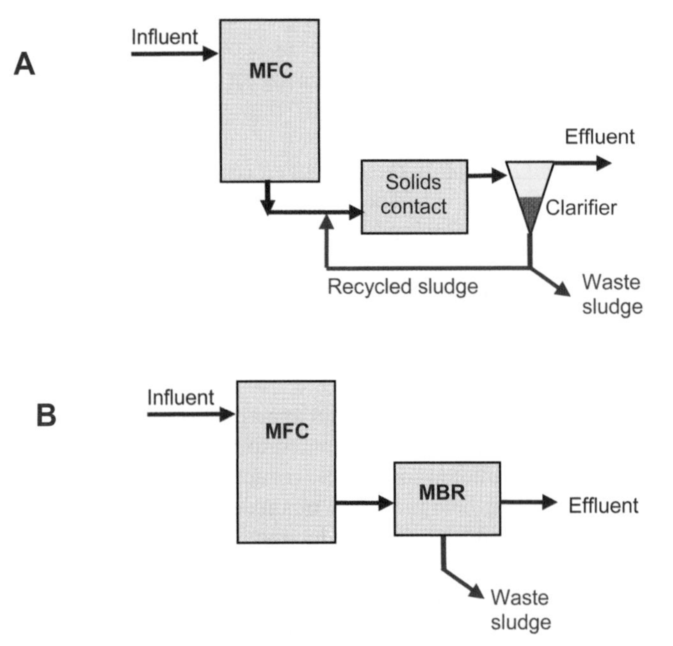

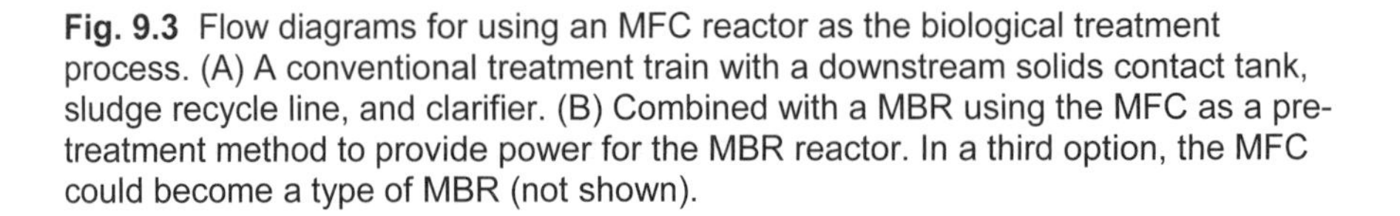
Fig. 9.3 Flow diagrams for using an MFC reactor as the biological treatment process. (A) A conventional treatment train with a downstream solids contact tank, sludge recycle line, and clarifier. (B) Combined with a MBR using the MFC as a pre-treatment method to provide power for the MBR reactor. In a third option, the MFC could become a type of MBR (not shown).

A second possibility for using the MFC in a process train is to have it as the pre-treatment process for a membrane bioreactor (MBR) process (Fig. 9.3B). The MBR process consists of membranes that pull the wastewater through the membrane, filtering out particles and leaving the biomass in the reactor. Thus, this reactor functions both as the treatment method and as the clarifier. The MBR process is expensive to operate due to the energy needed for aeration and filtration. The MFC could be used to offset those energy costs by providing electricity for the process, reducing solids production, and lowering the oxygen demand.

A third possibility for the MFC is to make it operate as an MBR. By creating cathodes that are able to function both as the cathode and filtration tube, the MFC could serve as a single MBR type of system. Work is ongoing on development of the cathode able to meet this criterion. The benefits of the combined system would be a smaller reactor and less energy needed compared to an MBR process, but it is unlikely that such a treatment plant could ever be a net producer of electricity. The main difficulty of operating such a system would be the potential for membrane clogging. MBRs are operated with aeration over the tubes to minimize fouling, but the MFC needs to be run as an anaerobic system. Intermittent aeration of the reactor could be performed as many electrogenic bacteria are facultative anaerobes, but further work is needed to more fully investigate this idea.

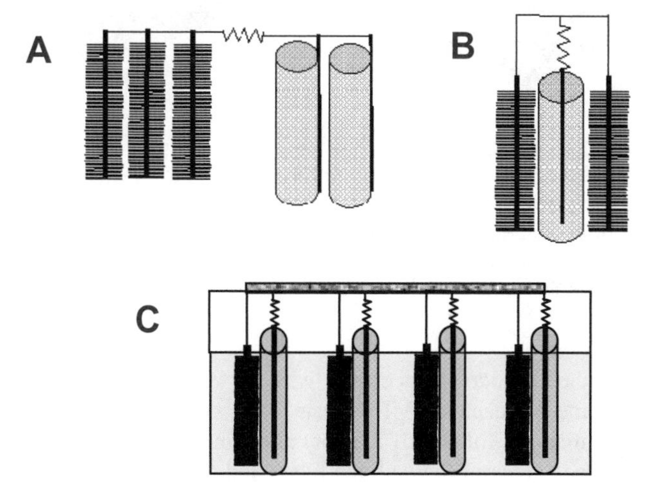

Fig. 9.4 Using brush anodes and tube cathodes. (A) Brush anodes arranged in series with cathodes, with conductive catalyst material (CCM). (B) anodes mixed in with tube cathodes. (C) Example of how the anode/cathode assemblies could be dropped into a tank, and used to treat wastewater (although resistors would not be used here). They can be linked in series to increase voltage (or in parallel to increase current).

Configuring the electrodes in the MFC. The MFC electrodes currently thought to be the most suitable for scale up consist of high specific surface area graphite brush electrodes and tubular cathodes. These electrodes could be placed in sequence (*i.e.*, anodes then cathodes) or randomly in the same tank (Fig 9.4). Groups of anodes and cathodes would form a single "anode-cathode module," with many electrodes placed in a single tank or module. The brushes and tubes within the module would need to be in close proximity to each other to minimize the distance for proton transfer and thus internal resistance (Fig. 9.4). Modules would then be linked together in the same manner that rotating biological contactor (RBC) modules are linked together for a treatment process where flow is directed through parallel trains, with each train consisting of several modules linked hydraulically in series. These modules could be electrically connected in series to boost voltage, or in parallel to boost current. The concept of placing many modules into a tank makes this process quite scalable for small or large systems. Like RBCs, additional tanks of modules could be added later on to increase treatment efficiency or flow capacity.

The cost of the tube cathodes are higher than those of the brush anodes on a surface-

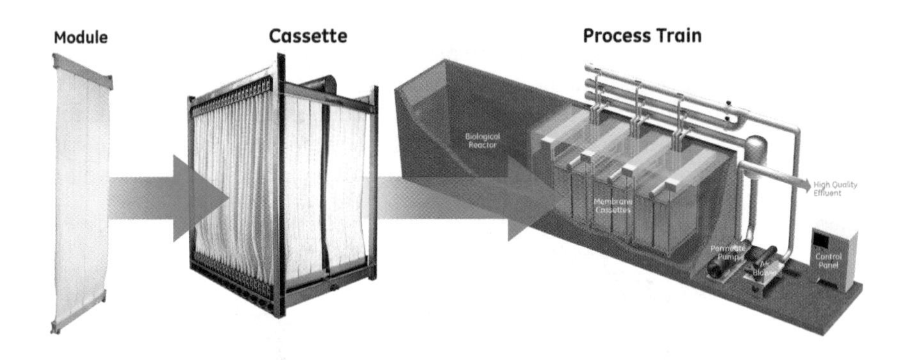

Fig. 9.5 Example of how hollow fiber membranes are used in wastewater treatment. In a membrane bioreactor (MBR), individual modules of membrane fibers are placed together to form a cassette, which is then integrated into a complete treatment system as shown here for the ZeeWeed 500 MBR system. (Diagram provided by J. Miro Chastven, reprinted with permission of GE Water and Process Technologies.)

area basis, and thus the cathode surface area will be the controlling factor in reactor performance. Membrane bioreactors (MBRs) can have specific surface areas of 180 to 6800 m^2/m^3 (surface area per volume of reactor) and can be built in modular form (Fig. 9.5). The relative surface areas of the cathode tubes and the graphite brushes are not yet optimized, and the costs for these materials cannot yet be reliably predicted. We believe costs for tube cathodes can be reduced to be less than those for inexpensive cation exchange membranes ($80/$m^2$). Certainly, high-cost materials like Nafion ($1400/$m^2$) can be avoided. Graphite fibers are relatively inexpensive, costing on the order of $0.60/$m^2$. It is likely that much more surface area will be used for anodes than for cathodes.

Estimating an HRT. Most MFCs developed to date have had HRTs of 24 h or longer and are operated in fed-batch rather than continuous mode. Systems with shorter HRTs are difficult for researchers to operate for practical reasons. While few studies have been conducted with continuous flow systems it should be possible to reduce the HRT to a range only slightly larger than that of an AS sludge reactor by optimization of the surface areas of the anodes and cathodes relative to the reactor volume. For example, using domestic wastewater (246 mg-COD/L) and a flat plate MFC, Min and Logan (2004) achieved 58% COD removal in a 2-h HRT, and 79% removal at a 4-h HRT, but they did not report on longer HRTs. These studies, like many others, were primarily designed to increase power production or study other aspects of MFC architecture, not to achieve high COD removal. AS aeration basins have HRTs typically in the range of 4 to 6 hours (*Tchobanoglous et al.* 2003; *Viessman and Hammer* 2005). Thus, a goal for an MFC system is to match this range of HRTs in order to produce a similar-sized system.

In order to provide a context for calculations made in this chapter, let us assume a WWTP treating a flow of 0.44 m^3/s (38,000 m^3/d; 10 million gallons per day = mgd), or the wastewater from 76,000 people in the US of 132 gal/d-person (500 L/d-person). For a 6-h HRT, this means that for the AS aeration stage we would need one tank 9500 m^3 (or several tanks with smaller volumes). Assuming that the influent wastewater has an average BOD_5 of 300 mg/L, the AS basin influent flow before recycle will be 200 mg/L of which approximately 1/3 is $sBOD_5$. We can assume the AS would produce an effluent of <20 mg/L BOD_5 using a suitable recycle flowrate.

For comparisons of MFCs to TFs, let us assume that the TF is loaded at 0.67 L/m^2-s (1 gpm/ft^2) based on the top cross-section of the tank. This would require a TF reactor 29 m (94 ft) in diameter, and based on calculations using the LTF model, a reactor height of 6.1 m (20 ft) tall (*Logan et al.* 1987; *Parker and Bratby* 2001) (or more likely 2 reactors each 20 m in diameter and 6.1 m in height), or a reactor volume of roughly 4000 m^3. The HRT of a TF is difficult to measure due to flow over the biofilm and diffusion of tracers into the biofilm. Typical reported HRTs for TFs start at 15 minutes for the retention of the peak concentration of the tracer, but longer times are more reasonable. The media has a typical porosity of 95%, but if we assume flow is through 20% of the void space we obtain a HRT for flow through the system of 0.5 h. If we were to operate the system as a completely saturated system, the HRT would be 2.4 h. To estimate BOD_5 removal for the TF, we can use the LTF model based on the TRIFIL model (*Logan* 1999; *Logan et al.* 1987). Experience has shown that $sBOD_5$ removal is predictable in a plastic media TF, and that pBOD removal occurs primarily as a result of the TF/SC and clarifier operation. Assuming cross flow media with a specific surface area of 100 m^2/m^3, the above loading rate and a $sBOD_5$ = 67 mg/L (with 20% recycle) in the influent, the overall removal is 77.3% producing a $sBOD_5$ effluent of 15 mg/L. Normalizing this to the reactor volume, the overall sBOD loading rate is 0.64 kg-$sBOD_5/m^3$-d, with a removal rate of 0.49 kg-$sBOD_5/m^3$-d, values which are reasonable for TF systems (*Viessman and Hammer* 2005).

The size of the SC process added on to a TF process can be estimated using a first-order model (*Matasci et al.* 1986), assuming a typical kinetic rate constant $K_{20} = 3 \times 10^{-5}$ L/mg-min. The effluent $sBOD_5$ concentration from the SC process (c_{eff}) relative to that influent which typically comes from the TF (c_{in}) is calculated as

$$\frac{c_{eff}}{c_{in}} = \exp\left(-K_{20}\,\theta^{(T-20)} X\,\tau\right) \tag{9-1}$$

where θ = 1.035 is a temperature correction factor, T (°C) is the wastewater temperature, X (mg/L) the volatile solids concentration, and τ (min) the HRT. Assuming the effluent of 45 mg/L of BOD_5 from the TF is to be reduced to 10 mg/L, T = 20°C and X = 2,000 mg/L, the required HRT would be 25 min. For a flow of 10 mgd, the SC tank size would be 660 m^3, or an additional 17% of tank volume compared to that needed for the TF.

We conclude from this comparison that for a 0.44 m^3/s flow an AS aeration basin size would be 9500 m^3, while that of the combined TF/SC would be 4,660 m^3. As stated above, it is believed that an MFC could be composed of a reactor approximately the size of the AS basin but needing an additional SC process. Adding in the volume of the SC tank, the size of the systems would be two tanks having a total volume of 1.02×10^4 m^3, with a 6-h HRT in the first tank and a 25 min HRT in the second tank. Pilot scale testing is needed to determine if such HRTs are possible, and if an MFC/SC process would work as suggested here.

Power densities based on substrate fluxes. The power that can be produced in an MFC is dependent on the substrate concentration, type of substrate, conversion efficiency, and application rate. Maximum power densities possible with simple substrates such as glucose based on mass transfer limitations are addressed in Chapter 7. For wastewater, the recoverable energy is not well studied. Shizas and Bagley (2004) estimated an energy content of 2.26×10^8 kJ/d for the North Toronto wastewater having a flowrate of 0.41 m3/s (35,700 m^3/d, 9.43 mgd). For the influent COD of 431 mg/L, this produces an energy content of 14.7 kJ/g-COD (3.5 kcal/g-COD). For comparison, the heats of combustion of glucose and acetate are 15.6 kJ/g and 14.6 kJ/mol.

To put this energy content of the wastewater into perspective based on a biofilm process in a TF, we can compare the substrate removal that is occurring in a TF to an equivalent power density based on the assumption that wastewater has an energy content of 14.7 kJ/g-sBOD. From the top to bottom of the tower 1 m^2 of applied cross-sectional area with a 6.1-m depth has 610 m^2 of total surface area, and therefore the substrate flux into the biofilm is 0.057 mg/m^2-s. Converting this substrate flux into energy, the overall power density would be ~0.83 W/m^2. This value of course is highly dependent on the energy content of the wastewater and Coulombic efficiencies.

In the above calculation, we based power on the surface area of the anode. However, power output per volume in an MFC may well depend on the cathode surface area as the anode area will probably be provided in excess using a graphite brush electrode. Let us assume that power densities per anode area above will translate to power densities normalized by the cathode (*i.e.,* the proton flux limits power, not electrons flux). Assuming that cathode surface areas will range from 100 to 500 m^2/m^3, then power output normalized to reactor volume would be 83 to 415 W/m^3. This lower range has already been achieved with air cathode MFCs in the laboratory (up to 115 W/m^3) (*Cheng and Logan* 2007), but so far only 18 W/m^3 have been obtained with tube cathode systems. Volumetric power densities above this range have reached 260 W/m^3 using ferricyanide (*Aelterman et al.* 2006), but with architectures that are not expected to be cost effective for scaleup. Thus, achieving these goals in power densities remains a challenge for MFC researchers.

9.3 Energy balances for WWTPs

Power used for wastewater treatment. The power required at different WWTPs varies widely, but general trends can be developed based on the type of treatment process. TFs are reported to use 430 kW per m^3/s, with AS using 2.4 to 5.9 times as much power (1020–2550 kW per m^3/s) (Table 9.1). These estimates for AS are a little higher than Shizas and Bagley (2004) reported for the North Toronto Wastewater Treatment (NT) plant of 680 kW per m^3/s (30 kW/mgd). The wastewater at this plant contained 431 mg-COD/L and 1930 mg-TS/L, and contained in the influent a total of 2616 kW. Thus, the energy content of the wastewater was 9.3 times that required to treat the wastewater (283 kW). A membrane bioreactor (MBR) system requires 8520 kW per m^3/s for treatment, and thus could not be self-sustaining based on the energy content of the wastewater.

Table 9.1 Energy usage for different wastewater treatment processes [from Cooper *et al.* (2007)]

Wastewater treatment process	Energy used, kW per m^3/s (kW/mgd)
Oxidation pond	170–430 (7.5–19)
Trickling filter (TF)	430 (19)
Activated sludge (AS)	1020–2550 (45–112)
+ nutrient control	840 (37)
Membrane bioreactor (MBR)	8520 (375)

Energy recovery scenario based on the NT plant. It is important to consider where the energy in the solids entering a plant move through the system so that we can determine how energy might be recovered from the wastewater. At the NT plant 66% of the energy content of the raw wastewater (2616 kW) ends up in the primary clarifier solids, leaving 34% of the energy contained in the primary clarifier effluent (886 kW) (Table 9.2). Of the original energy entering the plant 14% ends up in the secondary clarifier effluent, so that 80% of the energy in the original wastewater is captured in the sludge from the two clarifiers. At this plant 25% of the original energy exists in the treated solids (biosolids). It is estimated that 38% of the energy entering the plant is captured in biogas production as methane, and that based on energy content of the solids the biogas process captures 47% of the energy in the solids.

Table 9.2 Energy content of the North Toronto waste streams [data from Shizas and Bagley (2004)]

Stream	Flow (m^3/s)	Energy (kJ/d, 10^8)	Energy (%)
Raw wastewater	0.413	2.26	100
Primary effluent	0.409	0.759	34
Primary sludge	0.0036	1.50	66
Secondary sludge	0.0094	0.317	14
Treated solids	0.0515 (kg/s)	0.565	25
Biogas production	0.0397	0.858	38

These results have important implications for how an MFC type of wastewater treatment plant should be designed and operated. To maximize energy recovery as electricity, it may be best to omit the primary clarifier and send the raw sewage directly into the MFC reactor, although this might lead to reactor media clogging. This direct addition of wastewater without primary clarification would require that the MFC capture the energy in the solids over a reasonable HRT, a subject which has not been investigated yet as being feasible. The removal rate of solids in the primary clarifier at the NT plant is typical, although at the upper limit of that range. Usually ~1/3 of the BOD is removed in the primary clarifier and 60–65% of the total solids (*Viessman and Hammer* 2005), so here 2/3 we see 66% of the energy (total solids) is diverted away from the MFC and resides in the primary clarifier solids. Even with the remaining energy of 886 kW, however, there is still sufficient energy to make the MFC process self-sustaining. A reasonable goal for the MFC is to recover half the energy content of the wastewater at an energy input similar to that of a TF. An estimated 179 kW would be needed for an MFC (not including energy needed for a SC process) with 443 kW recovered (Fig. 9.6).

In this example for the NT plant, the MFC captures 17% of the energy as electricity while the anaerobic digesters capture 37% of the energy as methane, or a total of 54% energy recovery (Fig. 9.6). The NT plant uses AS for wastewater treatment, and thus use 680 kW versus the 179 kW assumed to be needed for the MFC, adding an additional 4% (104 kW/mgd) of energy savings, for an overall energy recovery of 58%. In this scenario the NT plant would consume 179 kW and produce 1518 kW, or be a net producer of 1.34 MW. Thus, a common wastewater treatment plant could be a net exporter of energy (electricity and biogas energy) rather than an energy consumer.

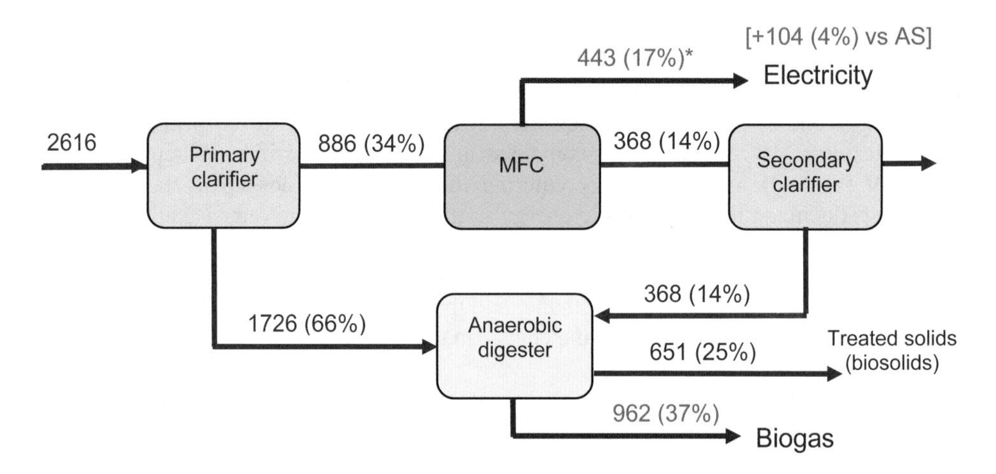

Fig. 9.6 Distribution of energy (kW; values in parentheses are percent of total entering in raw sewage) based on the energy balances at the NT plant [from *Shizas and Bagley* (2004)], except for the power production by the MFC which is estimated at 50% energy recovery (indicated with an asterisk). Note that the solids removal in the primary clarifier is higher than typical, and that less solids would be produced by an MFC than an AS process. Differences in energy flows into and out of a reactor are assumed to be dissipated as heat.

These calculations of energy production do not consider the conversion efficiencies for final electricity production. For example, a single MFC reactor will produce *ca.* 0.3–0.6 V which would need to be converted to a higher voltage (and AC) for use. This can be done by linking several MFCs in series and through DC-DC conversion or DC-AC conversion. For the AD reactor, the methane gas would be used to produce electricity, with the heat of gases produced used to heat the digesters. The conversion efficiency of methane to electricity varies with the type of system used, with values ranging from 30 to 40% for combustion systems and fuel cells (Table 9.3). Assuming 85% efficiency for the MFC voltage conversion and 35% for methane conversion efficiency, this represents a final electricity production of 717 kW (27% recovery) with another 104 kW (4%) saved versus an AS-based process, or a total of 821 kW (31%).

The process train shown in Fig. 9.6 shows that much of the influent energy is contained in the solids from the primary clarifier, and this energy is then captured as biogas through conventional anaerobic digestion. If these solids leaving the primary clarifier were fermented to soluble organics, then the organic matter could be used in the MFC. A process train that included fermentation of primary clarifier solids is shown in

Table 9.3 Conversion efficiencies, capital costs and nitrogen oxide emissions of methane biogas to electricity with different technologies (*Willis et al.* 2007).

Category	Rich-burn engine	Lean-burn engine	Advanced reciprocating engines	Microturbines	Fuel cells
Nitrogen oxide emissions (kg/MWh)	20	1.4	0.7	0.2	0
Net electrical efficiency	~30	~32	~40	~24	~40
Net electrical power (W/MBtu)	86	92	115	69	115
Capital cost ($/MWh)	$500k	$600k	$700k	$1.1M	$6.0M
Operating costs ($/MWh)	$12	$12	$14	$20	$20

Fig. 9.7, along with the solids contact process included in the process train for the effluent leaving the MFC. There are too many unknowns at this time to predict an energy balance for such a system, but presumably this process train would recover more energy directly as electricity and reduce the need for long detention time anaerobic digesters. The fate of the solids produced by the fermentation system, and those from the solids contact process, would need to be determined based on studying their characteristics in pilot scale tests of the proposed process train.

9.4 Implications for reduced sludge generation

Another advantage of the MFC process versus an aerobic treatment process, such as AS, is a reduced production of solids from the bioreactor. The subject of cell yields from MFCs is addressed in Chapter 7. As noted there, estimated cell yields from an MFC process with acetate are thought to be on the order of $Y_{X/s}$ = 0.16g-COD-cell/g-COD. This

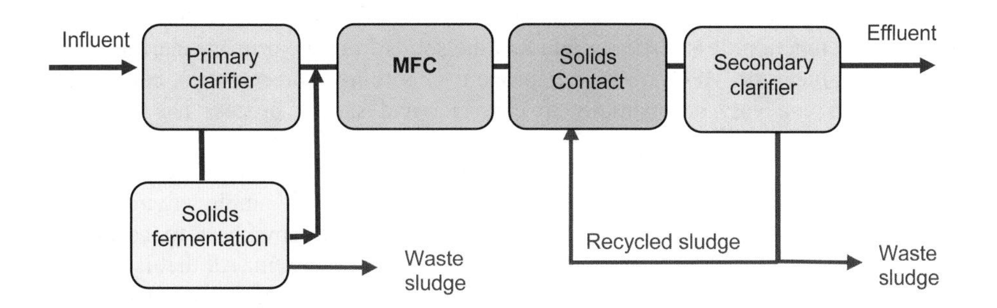

Fig. 9.7 Example of a process train for optimizing energy recovery in an MFC. If solids from the primary clarifier are fermented and included in the MFC influent, it would be possible to capture energy directly in the form of electricity rather than as biogas using anaerobic digestion. The fate of the waste sludge from these two processes would need to be evaluated after examining the characteristics of the sludge produced.

is about 40% of the value produced by an aerobic process of $Y_{X/s}$ = 0.40g-COD-cell/g-COD-WW. In the example above the production of solids from the AS system was 0.41 based on energy content of the solids. Thus, we expect that solids production from the MFC unit would be reduced to be only 40% of that generated from an aerobic process. This has the advantage of a substantial reduction in solids handling at a treatment plant, and thus a savings in capital expenditures. A solids contact (SC) process is also not included in these estimates of sludge production. The addition of an SC unit would age the solids, causing a further reduction in overall yield, but would also require additional energy input for reactor mixing.

A reduction in solids production from the secondary clarifier will reduce biogas production, which could adversely affect power production as methane, although this effect is expected to be offset compared to the potential savings for capital expenditures needed for the anaerobic digesters. In the example above, a 40% solids reduction from the secondary clarifier would reduce solids to 151 kW, and at the efficiency given above for biogas production result in biogas with an energy value of 858 kW (33% of energy in raw sewage). This is therefore a reduction of 4% of energy output from the digesters.

The settlability of the solids produced by the MFC is not known. The bacteria growing on the electrodes will form a biofilm, but the settling characteristics and the size spectra of such solids is not known. Bacteria will grow on the cathode, and we can expect that these biofilms will have characteristics similar to aerobic biofilms. However, these bacteria are sustained by oxygen diffusing through the cathode at the "bottom" of the biofilm, or at the support structure versus bacteria in a TF typically exposed to both oxygen and substrates at the "top" of the biofilm. The characteristics of the anode biofilm are also not studied. The bacteria on the anode are electrogenically active at the "bottom" of the biofilm where the conductive anode allows them to respire by transferring electrons to the electrode. As the bacteria at the electrode surface grow, they can be expected to push away older cells. Thus, the anode biofilm functions in a novel way in that the bacteria are most active at the biofilm-support surface, and they age as they get distant from the electrode. Thus, it is difficult to anticipate the characteristics of these biofilms for settling in a clarifier.

As mentioned above, a solids contact process will likely be needed to further remove BOD from the MFC effluent. This process could be expected to improve the settling characteristics of the MFC biofilms. It is unknown at this point how well the SC reactor system would function. It is also possible that the solids from the primary clarifier could be used to augment the MFC solids or improve their settling characteristics, but primary clarifier solids can vary substantially in character and such a process has yet to be examined.

It is clear that many questions remain on implementation of the MFC into a wastewater treatment system process train. However, there is ample environmental engineering experience to provide the needed judgment on how best to assess the performance of a new MFC system in a wastewater treatment system. All that is required now is the incentive and funding to begin such testing and analysis of these systems.

9.5 Nutrient removal

While BOD removal is one of the main goals of wastewater treatment, nutrient removal is also critical to the treatment objectives of most wastewater treatment plants. Little work has so far gone into following the fate of nitrogen and phosphorus in MFCs.

Nitrogen assimilation into bacterial biomass is an important nitrogen removal process, but it accounts for only a small percent of the total removal needed. In tests examining carbon balances around an MFC, Freguia *et al.* (2007) assumed nitrogen was incorporated into biomass with a stoichiometric formula of $CH_{1.75}O_{0.52}N_{0.18}$ where acetate was the electron donor. In a study on a cow-waste slurry, Yokoyama *et al.* (2006) measured 16% nitrogen removal and 30% phosphorus removal under conditions where 84% of the BOD was removed. In tests using swine wastewater (*Min et al.* 2005b) found 83 ± 4% ammonia removal with 86 ± 6% soluble COD removal over a 100-h batch cycle. Nitrate concentrations increased from 3.8 ± 1.2 to 7.5 ± 0.1 mg NO_3-N/L, indicating that nitrification due to oxygen diffusion into the single chamber MFC through the air-cathode.

In further tests Kim *et al.* (2007a) found 60% ammonia removal for a single–chamber MFC fed acetate with the cathode exposed to air (air-cathode) in five days. The main mechanism for removal was not due to biological processes, but rather to volatilization of ammonia at the cathode. During operation of the MFC the solution pH in the immediate vicinity of the cathode is raised, resulting of conversion of ammonium ion to ammonia which then volatilizes and is lost through the cathode to the air. This mechanism was demonstrated in abiotic tests, where it was shown that there was a nitrogen loss of 72% over 7 days at an applied voltage of 1.1 V. Ammonium ion was also lost from the anode chamber in two-chamber MFC tests, where ammonia decreased by 69% over 13 days using a ferricyanide catholyte, and by 89% over 20 days using dissolved oxygen at the cathode. Here the nitrogen loss was due to a combination of factors which included ammonium ion transport through the membrane by passive diffusion as well as by facilitated transport due to charge transfer.

It has also been found that nitrate reduced in the cathode chamber of an MFC, with bacteria accepting electrons from the cathode and transferring them to nitrate as the terminal electron acceptor (*Clauwaert et al.* 2007). An MFC was operated with both the anode and cathode reactions catalyzed by bacteria, making it possible to reduce nitrate to nitrogen gas. While this novel denitrification process might make it possible to use an MFC for nitrate removal, ammonia would still first need to be oxidized to nitrate in order for this system to function. Ppwer densities produced using this biological denitrification system were relatively low compared to those obtained with oxygen, reducing the usefulness of the process from the perspective of a method for power generation.

These findings on nitrogen removal show hold great promise for new types of nutrient removal processes for WWTP, particularly if nitrogen can be removed without the need for additional post-treatment carbon sources or through water recycling. Clearly, additional research is needed on this important topic of nutrient removal for wastewater treatment.

9.6 Electrogenesis versus methanogensis

Most of the published work on MFCs has considered power production from soluble substrates. However, particulate substrates can serve as substrates for electricity production in an MFC. Recent studies have shown power generation in MFCs from chitin and cellulose (*Niessen et al.* 2005; *Ren et al.* 2007; *Rezaei et al.* 2007), and there are new studies being conducted on power generation from sludges and the removal efficiencies of MFCs for particulates in wastewaters. Thus, there will soon be much more data available on the potential for power production from particulate materials. Does this mean that there is a possibility of replacing anaerobic digesters with MFC-based systems?

AD with methane production requires the hydrolysis of particulate substrates into soluble substrates, followed by reduction of these to volatile acids and hydrogen that ultimately can be used by methanogens for methane generation. Two-stage digesters have been built that separate the hydrolysis/fermentation stages from the methanogenesis stage. Digesters are also often plagued by buildup of volatile acids and a low pH that inhibits methane production, leading to "stuck" digesters. It is possible that an MFC could be used for electrogenic digestion whereby solids are similarly converted to volatile acids but then these are used for electricity generation by exoelectrogens and not methanogens.

The rates of hydrolysis and fermentation by bacteria are much greater than those for methane generation, and so this step in sludge (particle) breakdown can occur in a reactor with a low HRT. In a second stage it might be possible to run an MFC, with the faster growth rates of exoelectrogenic bacteria allowing them to out-compete methanogens and resulting in an electrogenic reactor rather than a methanogenic process. We do not have sufficient information on this idea. It is probable that in the main MFC the hydrolysis rate may be too low to allow sufficient particle degradation in an MFC (assuming the HRT is 10 h or less) without solids recycle. Thus, a separate solids handling system would likely need to be developed if pBOD degradation was of interest. Even with such a solids system based on electrogensis, it is not clear whether exoelectrogens could outcompete methanogens under anoxic conditions, and whether volumetric power densities would be sufficient to allow this technology to economically compete with anaerobic digesters.

Comparison on volumetric power. MFCs can be compared with anaerobic digestion (AD) on the basis of volumetric power. The power produced in an anaerobic digester from 1 kg of COD is approximately 1 kWh, with a volumetric power density in practice of about 400 W/m^3 (*Pham et al.* 2006). MFC power densities have reached 115 W/m^3 using flat air-cathodes (*Cheng and Logan* 2007), and thus on a volumetric basis the power generation needs to be improved by about a factor of 3.5 to be comparable to anaerobic digestion. Continued progress in power output has been achieved with MFCs, so this may be possible (*Logan and Regan* 2006a). As outlined above, the goal for MFC power densities is in the range of 83 to 415 W/m^3. If the product gas from AD is to be immediately used, no storage of the gas is needed; otherwise, gas storage may be necessary adding additional costs. The MFC also has the advantage of directly producing electricity whereas the AD requires a separate co-generation plant that can give off air pollutants such as nitrous oxides (Table 9.3).

Comparison based on capital costs. Another way to compare AD with MFCs is on the capital costs. Sludge treatment can cost €500 ($650) per ton (*Rabaey and Verstraete* 2005), and one ton of carbohydrate treated with AD can produce methane that could produce useful energy of €160 ($210) (*Pham et al.* 2006). The capital costs for AD are on the order of €100,000 per ton of COD treated per day (*Pham et al.* 2006). If we view the AD process as strictly a power generation process, this means that it costs €2400/kW, or $1800/kW. In the power industry, capital costs are generally considered to be around $1000/kW (*Grant* 2003). Considering the value of the power combined with the costs, this indicates from this very simple approach that energy production does not offset capital costs.

The capital costs for an MFC are not known. However, we can begin to put some goals on costs based on this analysis. For example, TF media costs $15/ft^3 installed, or $530/m^3. If we achieved the same power output with a TF as that with AD (400 W/m^3), then we could produce electricity at \$1300/kW. If we achieved only 100 W/m^3, this would fall to \$5200/kW. To make the capital costs of the MFC comparable to that of the AD process, we would therefore need to achieve a capital cost of \$720/m^3 for the installed MFC components. To create a process that just produced power (*i.e.,* to achieve \$1000/kW), we would need to achieve \$400/m^3, which does not appear to be likely given the costs for TF media.

Other performance considerations. The main advantages of MFCs over AD will likely be temperature range, COD loading, and stability of performance. MFCs have been shown to function over thermophilic, mesophilic, and psychrophilic temperature ranges (*Holmes et al.* 2004c; *Jong et al.* 2006; *Liu et al.* 2005a, b; *Liu and Logan* 2004). Rates in MFCs between 20°C and 30°C change by very little (~10%) (*Liu et al.* 2005a), but AD is appreciably slower at temperatures below 30°C and it is inefficient at 20°C for the relatively dilute wastewaters that are typical of most domestic and some industrial wastewaters in the US. While MFC power production is improved by warmer temperatures, one study found that reducing the temperature form 32°C to 20°C reduced MFC performance by only 9% (*Liu et al.* 2005a). Thus, there is a good temperature tolerance of MFCs for power generation. MFC performance over a wider temperature range has not been examined in a single reactor system.

MFCs work over a wide range of COD loadings. In contrast, AD is only suited for high COD loadings as part of the energy produced as methane gas is needed to keep the reactors running at optimal temperatures (36°C). Thus, dilute wastewaters (such as domestic wastewater) will not have sufficient heat content to warm the wastewater and allow treatment. This means that the role of AD is restricted to sludge treatment and certain industrial wastewaters where the process performs well and produces excess energy. The ability of MFCs to treat concentrated wastewaters has been shown using animal (swine) wastewaters (*Min et al.* 2005b), but HRTs and optimum conditions for treatment have not been sufficiently investigated to know if MFCs would be competitive for high COD wastewaters.

The greatest advantage of the MFC compared to AD may turn out to be stability and performance. The start up times for AD processes can be quite long (months) compared to MFCs which quickly can generate stable power output in as little as a few days (*Liu et al.* 2005a, b; *Liu and Logan* 2004). AD processes can fail due to buildup of volatile acids, resulting in low pH values. The same conditions could lead to failure of an MFC, but so far the operation of the MFCs look quite stable. However, most investigations of MFCs have been conducted using buffered solutions, so the comparison of this situation may not be equal in many cases. MFCs do require biofilms grown on support media and cathodes for oxygen reduction, and in some cases they may contain membranes which could potentially foul. There remains much work to be done examining the longevity of MFCs in terms of stability of power output and long term performance.

Given these differences in performance for MFCs and AD, it seems likely that in the near future the two processes may function best as complimentary processes, with the MFC used for dilute wastewater treatment and ADs used for sludge treatment. This separation of processes may change over time, however, as MFCs are improved and applied over a wider range of conditions.

CHAPTER 10

Other MFC Technologies

10.1 Different applications for MFC-based technologies

Until now discussion has primarily centered on the use of MFCs for either wastewater treatment or as a method of renewable energy production in the form of electricity or hydrogen. However, MFCs have been examined for two other applications that will be examined in this chapter. The first one is using the MFC as a remote source of power. Reimers *et al.* (2001) first developed this new concept of a sediment MFC (SMFC) and showed that power generation could be sustained by bacteria using only the organic matter in the sediment. Since then, there have been improvements in power output through modification of the materials, but also in the development of enhanced SMFCs through augmentation of the anode with additional sources of energy. The second novel application of the MFC is as a method of bioremediation. Remediation can include both degradation of organic pollutants at the anode as well as reduction of inorganic chemicals such as nitrate or uranium at the cathode.

10.2 Sediment MFC

In the late 1990s, scientists were hotly debating how iron reducing bacteria such as *Shewanella* spp. and *Geobacter* spp. could transfer electrons from intracellular respiratory enzymes to outside the cell (and as discussed in Chapter 2 the debate continues!). Nobody questioned that electrons were being transferred exogenously, it was just that the mechanism(s) were unconfirmed. Reimers *et al.* (2001) decided to put this ability of extracellular electron transfer to work. In 2001, they published a paper on harvesting the energy contained in marine sediments in a completely new type of MFC. While others had shown that electricity could be produced with mediators for quite some time (*Potter* 1911), and Kim and coworkers had already patented the idea of using an MFC for electricity production (*Kim et al.* 1999c), this work had mostly gone unnoticed by many scientists. Reimers *et al.* (2001) were able to highlight to a wide scientific audience that bacterial processes could be harnessed to produce useful amounts of energy, in this case for producing power in remote and relatively inaccessible areas such as the seafloor. The SMFC was used as a power source for devices place on the seafloor, but it is also possible it could serve as a type of "refueling station" for small autonomous devices that are operated to the underwater environment (Fig. 10.1).

The concept of the SMFC is quite simple: Place the anode into the anaerobic sediment and place the cathode into the overlying water containing dissolved oxygen. The high salinity of the seawater provides good ion conductivity between the electrodes, and the organic matter needed by bacteria to produce electricity is already in the sediments. The remarkable part of the tests at that time was that exoelectrogenic bacteria were already present in the sediments, and that they were able to sufficiently compete with microorganisms using other electron acceptors to produce useful amounts of power. They noted that sediments on continental margins typically contained 2–3% of organic carbon (dry weight) and that the constant flux of new carbon into sediments if captured by such devices could sustain power generation at a level of 50 mW/m^2 essentially indefinitely.

Fig. 10.1 Concept of using a sediment MFC as a remote power source on the seafloor either for powering a data collection device or serving a "refueling" station. (Graphic by R. Brennan.)

Initial tests by Reimers *et al.* (2001) using all Pt anodes and cathodes produced power densities as high as 15 mW/m^2. They also conducted tests using carbon fibers sandwiched between fiberglass screens and plexiglass frames, and cathodes, and produced initially up to 42 mW/m^2, with power dropping over the next 80 days to 10 mW/m^2. Power was sustained at a level of 4–10 mW/m^2 in tests conducted over the next 160 days, for a total of 240 days. They hypothesized that organic matter, HS^-, and Fe^{2+} oxidation could all be sources of energy for power generation, oxygen reduction was occurring via water or H_2O_2 formation, and humics present in the sediments could be acting as electron mediators.

Field tests of SMFCs. Tests with large SMFCs were conducted by Tender *et al.* (2002) at two sites, one in Yaquina Bay Estuary near Newport, Oregon, containing 2–6% carbon (dry weight), and one in a salt marsh area near Tuckerton, NJ with sediments containing 4–6% carbon. The SMFCs were constructed from graphite discs 48.3 cm in diameter and 1.3 cm thick, perforated with 790 holes evenly spaced at distances of 0.64 cm, producing a total projected surface area of 0.183 m^2 (one side only). The area of the electrode based on both sides, and using area exposed by the holes, was reported by others to be 0.542 m^2 (*Ryckelynck et al.* 2005). At the Tuckerton site the electrodes were enclosed within 167 L PVC containers drilled with 80 holes, each 5.1 cm in diameter to admit seawater (Fig. 10.2). Power density curves showed that the site at Newport produced 32 mW/m^2 while 18 mW/m^2 was produced at the Tuckerton site. A continuous power density of 28 mW/m^2 was achieved at the Newport site. The Tuckerton site was used to examine power by the SMFCs under different current and voltage conditions, but the earliest continuous power densities achieved a maximum of 27 mW/m^2. Slopes taken from polarization curves imply internal resistances in these two SMFCs of 21 and 24 Ω. The bacteria that were enriched on the anodes in these reactors were discussed in Chapter 2, consisting primarily of microorganisms in the family *Geobacteraceae*.

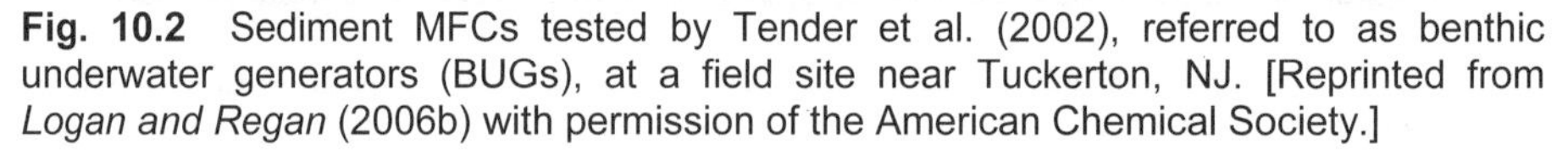

Fig. 10.2 Sediment MFCs tested by Tender et al. (2002), referred to as benthic underwater generators (BUGs), at a field site near Tuckerton, NJ. [Reprinted from *Logan and Regan* (2006b) with permission of the American Chemical Society.]

The same type of SMFC was tested by Ryckelynck *et al.* (2005) in Yaquina Bay, OR, with the device located at water depth of 5 m (low tide). The electrodes were the same as used by Tender *et al.* (2002), except here the available surface area was defined based on all exposed surfaces (0.542 m^2). The reactor produced continuous power at 11 mW/m^2. If the projected surface area is instead used, this is equivalent to 33 W/m^2, which is a range similar to that reported by Tender *et al.* (2002) for this device at the Newport location. Iron and sulfur were enriched on the anode 400 and 20 times, respectively, after operation at this site. Community analysis of the bacteria showed a predominance of δ-Proteobacteria, primary sulfate reducing bacteria consisting of *Desulfobulbus* spp. and *Desulfocapsa* spp. It was concluded that sulfides (primarily FeS and FeS_2) were the primary chemical sources used by bacteria for power generation.

Reimers *et al.* (2006) conducted field tests with a new type of SMFC consisting of a single graphite rod anode (8.4 cm diameter, 91.4 cm long) imbedded in the sediment at a deep-ocean cold seep, along with 1-m long cathodes made of carbon-fiber/titanium wire brush cathodes. The cold seep was located in Monterey Bay, CA, at a depth of 957 m. The reactor was placed there using the remote oceanographic vehicle (ROV) Ventana. The reactor produced a maximum sustained power output of 34 mW/m^2 of anode surface area over the first 26 days of operation. During the next 72 days, power decreased to less than 6 mW/m^2. Analysis of one of the anodes after recovering of the systems showed that there was extensive and uneven sulfur deposition, leading to the reduced performance of this system over time.

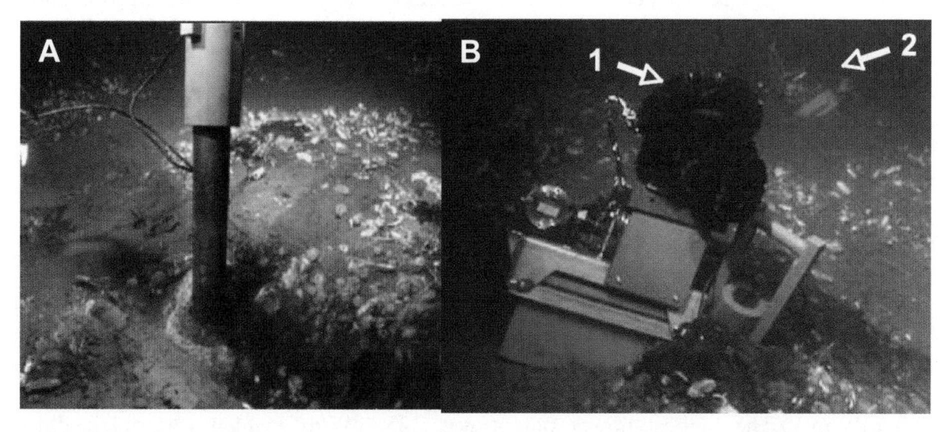

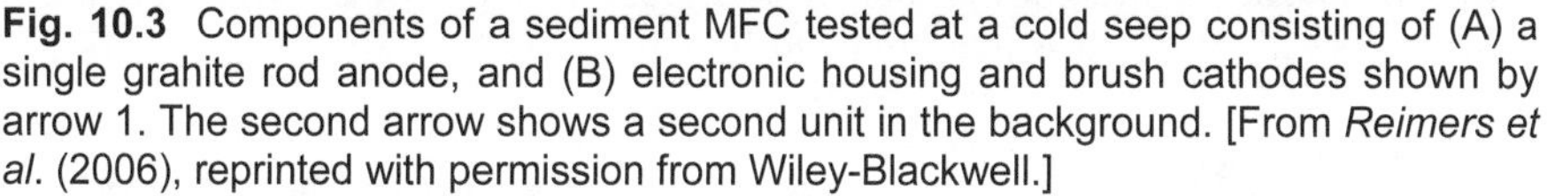

Fig. 10.3 Components of a sediment MFC tested at a cold seep consisting of (A) a single grahite rod anode, and (B) electronic housing and brush cathodes shown by arrow 1. The second arrow shows a second unit in the background. [From *Reimers et al.* (2006), reprinted with permission from Wiley-Blackwell.]

The cathodes used by Reimers *et al.* (2006) were modeled after the brush cathodes developed by others (*Hasvold et al.* 1997) for oxygen reduction (Fig. 10.3). They developed seawater batteries based on using sacrificial metal anodes made of magnesium and cathodes made of carbon fibers. The carbon fiber cathodes were noted to be superior in performances to stainless steel wool anodes that were previously examined. Their system used seawater as the electrolyte and the battery produced 650 kWh of power. A DC/DC converter system increased the voltage produced by the batter from 1.6 V to 2 W, with a 56% efficiency, resulting in the production of 2W for a prototype. A full scale system produced 30 W continuously with a maximum of 40 W at 1.5 V, with an improved DC/DC conversion efficiency of around 70–75%. The same type of battery was later used to power an unmanned underwater vehicle, but this time they used an aluminum anode and continuously recirculated an alkaline electrolyte containing hydrogen peroxide (*Hasvold et al.* 1999). The four-battery system produced 600 W at 30 V.

Sulfides are an important source of power for SMFCs. Using a two-chamber MFC, Rabaey *et al.* (2006) showed that power could be generated and sustained using a medium containing inorganic sulfides and no organic carbon. The MFC contained graphite granules in the anode chamber, and ferricyanide was used as the catholyte. The reactor produced 20 W/m^3 (37 W/m^3 based on anode liquid volume). It was also shown using a tubular type of reactor (*Rabaey et al.* 2005b) that power could be generated from glucose or acetate and sulfides when both were fed to the reactor (*Rabaey et al.* 2006).

Thus, sulfides can serve as a source of power in both the absence and presence of carbon sources.

Modifications of seawater electrodes to increase power. The power produced by SMFCs can be increased using the same principles developed to increase power in MFCs lacking a membrane. The high conductivity of seawater is an advantage compared to typical MFC systems that operate at much lower conductivities, but SMFCs must rely on dissolved oxygen at the cathode. Thus, these systems are designed with very large cathodes. To increase the performance of the anodes, they can be modified to contain mediators (*e.g.*, AQDS and NQ). SMFCs with these mediators produced 1.5 to 1.7 times as much power as systems with plain graphite anodes (see Chapter 5) (*Lowy et al.* 2006). Adding Mn^{2+} to graphite electrodes in field-tested SMFCs produced 105 mW/m^2, compared to 20–40 mW/m^2 achieved in previous tests. The main limiting factor in these systems likely is the electrode spacing, as discussed above, which results in high internal resistances in these systems.

Bergel *et al.* (2005) discovered that the formation of seawater improved the performance of stainless steel cathodes used in seawater applications of hydrogen fuel cells. The amount of improvement depended on the electrode sizes and pH. The fuel cell produced 41 mW/m^2 in the presence of the biofilm, but only 1.4 mW/m^2 when the biofilm was removed, an increase of 30× (power normalized to cathode projected surface area). Increasing the pH of the anode compartment from 8.2 to 12.5 increased power to 270 mW/m^2, which was almost 100× larger than that with the cathode biofilm removed (2.8 mW/m^2). The highest power output was 325 mW/m^2, which was obtained by decreasing the cathode surface area from 9 to 1.8 cm^2.

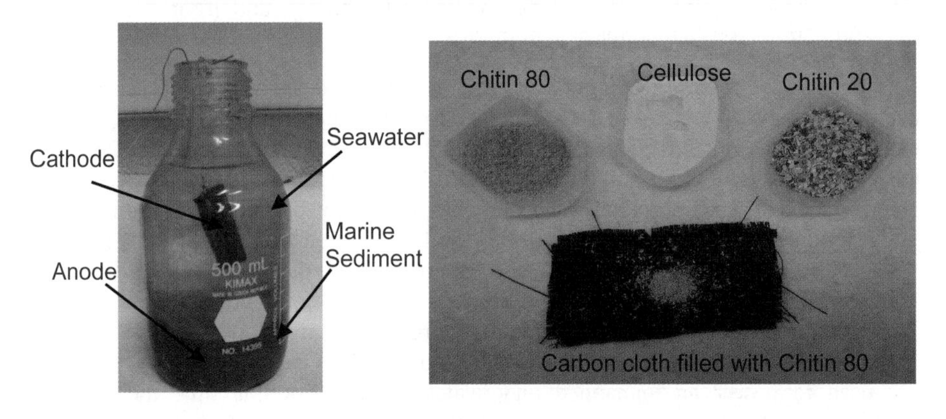

Fig. 10.4 (A) Substrate enhanced MFC (SEM) tested in the laboratory (aeration of the water above the sediment is not shown); and (B) example of the carbon cloth anode before being sewn to contain one of three different substrates shown. [Reprinted from *Rezaei et al.* (2007) with permission of the American Chemical Society.]

10.3 Enhanced sediment MFCs

Power generation by SMFCs is limited by the concentration of organic matter in sediments, and by the rate at which this material can be degraded by bacteria.

Recognizing these limitations, Rezaei *et al.* (2007) examined a different approach for powering SMFCs: incorporating biodegradable fuel, in the form of particulates substrates, into the anode assembly. They referred to this as a substrate-enhanced sediment MFC (SEM), although the added fuel is only supplied once. Thus, it acts more like a biobattery than a fuel cell as the power output is eventually reduced to whatever low level the sediment can support once the particulate substrate is exhausted. Tests were conducted in laboratory SEMs using two different types of chitin or cellulose as substrates. The SEMs were media bottles filled with sediment and water, with the cathode placed in the overlying water which was aerated using a stone diffuser (Fig. 10.4). Chitin is a polysaccharide of N-acetyl-β-D-glucosamine ($C_8H_{13}NO_5$). It is the second most abundant material in the world after lignocellulose (*Ryu and Nam* 2000), and it is readily degraded in marine systems as it is composed primarily of crab shells. For example, ChitoRem™ SC-20 (Chitin 20; JRW Bioremediation, LLC, Lenexa, KS) made from crushed crab shells contains on average approximately 20% chitin, 25% protein, 40% calcium carbonate, and 15% water. ChitoRem™ SC-80 (Chitin 80), made from deproteinized and demineralized crab shells, contains approximately 95% chitin, and is sieved to pass a #20 mesh (< 841 μm). The higher protein content of Chitin 20 makes it more biodegradable than Chitin 80. Celluose particles with a mean size of 50 μm were also examined as a biodegradable fuel. Anodes for all tests were prepared from carbon cloth sewn into a pillow-shape and then filled with the particles. These were immersed in marine sediment with the cathodes suspended in the overlying water.

Maximum power densities using chitin in the SEM were 76 ± 25 mW/m^2 and 84 ± 10 mW/m^2 (normalized to cathode projected surface area) for Chitin 20 and Chitin 80, respectively, versus less than 2 mW/m^2 for an unamended control. Over a 10-day period, power generation averaged 64 ± 27 mW/m^2 (Chitin 20) and 76 ± 15 mW/m^2 (Chitin 80). When cellulose was used, a similar maximum power was initially generated (83 ± 3 mW/m^2), but power rapidly decreased after only 20 hours. Power production varied among reactors set up in the same way, and it was thought that differences between reactors were due to a combination of several factors including different start-up rates caused by differences in bacterial access to the substrate, variations among reactors in particle sizes or biodegradability, and the performance of the cathodes (due both to local dissolved oxygen concentrations and performance on a comparable area basis). The internal resistances of these systems were 650 ± 130 Ω and 1300 ± 440 Ω for the Chitin 80 and Chitin 20 substrates and 1800 ± 900 Ω for cellulose. These values are all quite high relative to those estimated for the larger scale system tested by Tender *et al.* (2002). Thus, it is likely greater power densities could be achieved with more efficient SMFC designs.

The use of slowly degraded substrates in particulate form presents the possibility of both increasing power densities through greater availability of the substrate, as well as duration of the increased power through control of particle sizes. Surface area can limit the rate of biodegradation of the particles, and thus increasing particle size (and therefore controlling particle surface areas) should extend longevity of the system in a predictable way. A preliminary model suggested that longevity of power production would increase with particle diameter squared. Additional work is needed to test the concept and performance of SEMs in the laboratory, particularly with respect to particle size, as well as in actual field situations.

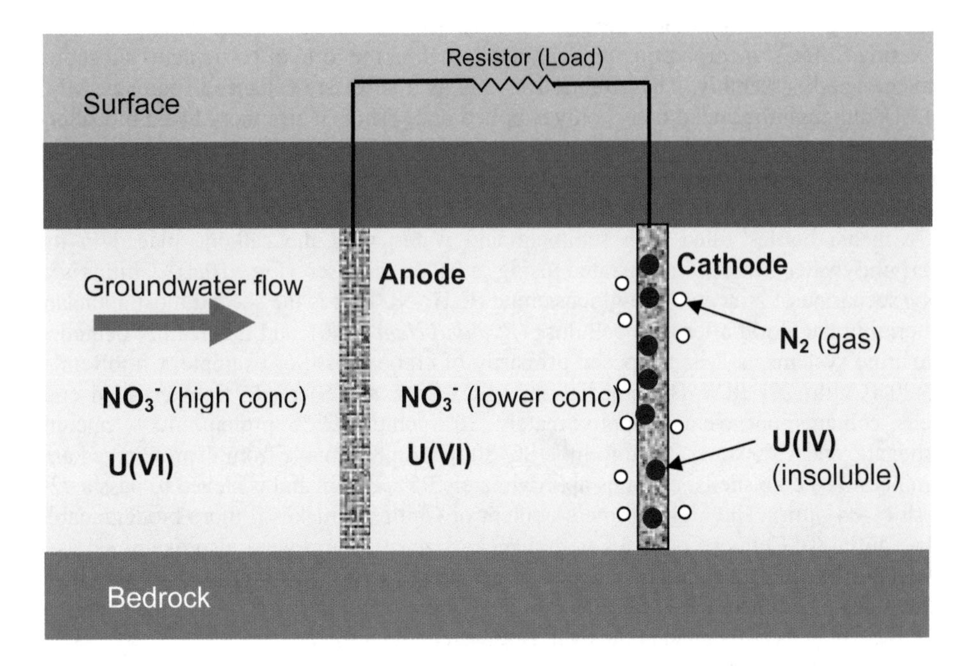

Fig. 10.5 Schematic of a potential method for using MFCs to remove soluble nitrate and uranium (VI). The anode is a conductive, non-corrosive carbon material that is wrapped around a particulate substrate such as chitin that is slowly degraded by bacteria. Leakage of organics and result in bacterial nitrate and U(VI) reduction in the vicinity of the anode when bacteria use them as the electron acceptor, resulting in NO_2^-, N_2 gas, and U(IV). Biodegradation of chitin and intermediate degradation products at the anode generate current. As the water reaches the cathode, the remaining nitrate and U(VI) can be further reduced by receiving electrons from the cathode. The cathode can be extracted for permanent removal of the U(IV). If needed, the potential generated by the circuit can be augmented with additional voltage in the circuit [as in the MEC process of *Liu et al.* (2005c)].

10.4 Bioremediation using MFC technologies

In an MFC the bacteria donate electrons to the anode using mediators or nanowires, but it is also appears true that that the transfer of electrons is reversible, *i.e.,* that bacteria can accept electrons from an electrode. This is the basis of the idea of a biocathode, which was briefly mentioned in Section 6.7, and is reviewed elsewhere (*He and Angenent* 2006). A biofilm catalyzing oxygen reduction is likely the basis for improvement in cathode performance observed by Bergel *et al.* (2005), described above. The concept of in-situ bioremediation is shown in Fig. 10.5. In this conceptual approach to groundwater remediation, the anode is powered by a long-lasting particulate substrate (such as chitin) and used to produce electricity. The electrons are then used by bacteria at the cathode in this example to accomplish either nitrate or U(VI) reduction.

It is not clear when the first applications of biocathodes occurred in MFCs. Reactors had been designed for electrochemical catalysis of organic and inorganic chemicals (*Li and Farrell* 2000; *Moraes and Bertazzoli* 2005), but such reactors are not considered here as they are not bacteria-based systems but rather pure electrochemical systems. In 1999, Matsumoto *et al.* (1999) showed that logarithmic growth of an iron-oxidizing bacterium, *Thiobacillus ferrooxidans*, could be extended from a final concentration of 4×10^9 cells/mL to 10^{10} cells/mL over an additional three days by introducing a Pt electrode into the reactor with an electrode set at ~0.2 V vs. NHE (0 V vs. Ag/AgCl). It was thought that potentiostatic electrolysis removed Fe(II), thereby achieving continued growth by reducing product inhibition, but the possibility of these bacteria directly accepting electrons was not considered.

The idea of using biocathodes for in-situ nitrate bioremediation was explored by Gregory *et al.* (2004) using mixed cultures and a pure culture of *Geobacter metallireducens* in a completely anoxic system. *Geobacter* species were known to produce electricity in an MFC (*Bond et al.* 2002; *Bond and Lovley* 2003) and to use nitrate as an electron acceptor, but the idea of using an electrode to donate electrons to bacteria had not been explored. With an applied potential of ~0.7 V (vs. NHE; 0.5 V vs. Ag/AgCl) pure cultures of *G. metallireducens* used electrons from the working electrode (cathode) to convert nitrate to nitrite. River sediments used in the same system similarly converted nitrate to nitrite only when supplied electrons using a poised cathode potential. It was also observed that cultures of *G. sulfurreducens* were able to reduce fumarate to succinate. The ability of these bacteria to accomplish nitrate reduction is a meaningful step towards remediation. In this case, however, nitrite in groundwater is less desirable than nitrate and thus additional work would be needed to find a way to accomplish complete denitrification to nitrogen gas. In addition, the voltage applied here is higher than is likely to be achieved in a biocathode system sustained by bacterial oxidation of organic matter at the electrode.

The same poised cathode approach was used by Gregory and Lovley (2005) to achieve reduction of soluble U(VI) to insoluble U(IV). When a potential of −0.5 V was applied in the absence of bacteria U(VI) was removed from solution but not reduced as the uranium re-dissolved into solution when the potential was removed. When *G. sulfurreducens* was added to the cathode chamber U(IV) was removed from solution and 87% was recovered on the electrode. U(IV) was not observed in the solution after 600 h when the system was maintained under anoxic conditions. However, if exposed to oxygen the uranium would be expected to rapidly dissolve. This scenario for U(VI) reduction presents the possibility of using this system to remove uranium from contaminated groundwater by concentrating it on the electrode and then removing the electrode from the system. The uranium could then be concentrated in an external solution, and the electrode re-used in the system. While this would work in a fully anoxic condition, any intrusion of dissolved oxygen into the system would be a problem as this could cause re-dissolution of the uranium, potentially creating a spike in chemical concentration.

Recent tests have also shown that perchlorate can be electrochemically stimulated, but it is not clear whether perchlorate reduction was accomplished through electron transfer into the bacteria or by bacteria growing on hydrogen evolved as a result of water electrolysis (*Thrash et al.* 2007). Initially, perchlorate reduction required the addition of a mediator (AQDS) but later it was found that an isolate could achieve perchlorate reduction in the absence of the mediator. However, there were low concentrations of hydrogen at the cathode surface. Recent studies have found that zero valent iron can stimulate perchlorate reduction by hydrogen oxidizing bacteria, and autotrophic bacteria have been isolated that can grow using hydrogen and perchlorate (*Yu et al.* 2006; *Zhang et al.* 2002). Thus, additional work will be needed to determine the method used by these bacteria to degrade perchlorate under these electrochemically assisted conditions.

Bioremediation at the anode. In the above examples, chemicals were reduced at the cathode. However, it may also be possible to accomplish chemical oxidation at the anode in situations where there is a high concentration of biodegradable organics. This would require sufficient electron acceptor being present at the cathode. For example, at a site contaminated with petroleum or gasoline the water could be directed through two consecutive hydraulically conductive chambers, much in the same manner that walls of zero valent iron are used to treat chlorinated aliphatics in groundwater. The first section would contain the anode, such as graphite granules, where the chemical would be oxidized (assuming anaerobic conditions) on the anode providing current to the cathode. The second section would contain a cathode, for example tube cathodes, where oxygen would serve as the electron acceptor. Leakage of oxygen into the groundwater would have an advantage of providing additional electron acceptor into the water to either allow continued treatment or to increase the oxygen concentration in the groundwater. This concept was recently explored using a single chamber, air-cathode MFC, with petroleum-contaminated groundwater (dominated by diesel range compounds) (*Jin and Morris* 2007). Sustained power (as high as 120 mW/m^2 cathode) was generated from the biodegradation of petroleum compounds. The rates of petroleum compound degradation were significantly increased compared to background rates. This preliminary study suggests that MFC types of systems could be used to enhance bioremediation of petroleum contaminanted groundwaters under anaerobic conditions. Additional studies are needed in this area.

CHAPTER 11

Fun!

11.1 MFCs for new scientists and inventors

Many scientists and engineers begin their careers as young inventors, first building volcanoes or collecting insects for school science fairs, then moving on to more advanced studies and experiments that require more fundamental science and math foundations. At the heart of all these investigations is the desire to learn what makes things work and, if we are really lucky, to invent something new that will benefit mankind (and getting rich would be okay too!). Inventors can be young or old! Continuing to invent throughout your life—thinking of new ideas and ways to improve existing designs—will keep you young and your mind agile.

MFCs are a perfect platform for discovery, offering the inventor a chance to simultaneously delve into biology, chemistry, electronics, and physics. The choices that are made right from the start should reflect which of these aspects of design will be more exciting for you to focus on in your tests. Begin by asking questions about what things you will need to get started. Where will the inoculum come from, and what bacteria will develop? What chemicals will be used for the solutions, and what materials for the system architecture will work best? What does power really mean, and how much do you want to produce? What materials are readily available and what will your design cost to build? The cost will especially drive your design as you will probably want to make use of inexpensive materials that you can readily buy off the internet or from a local store. But don't be discouraged by possible costs! All engineers are challenged by the task of making something as good as possible, as quickly as possible, for as little money as possible (the so-called "better, faster, cheaper" motto).

As you make your choices and design your experiment, remember the scientific method: form a hypothesis, design an experiment to test that hypothesis, and then test it. Did you prove your hypothesis? What else did you find out, and what new hypotheses will you investigate next? Remember as you design your experiment that it is best to tackle only one aspect of design or operation at a time. While more advanced experimental design will allow you to test several variables (factorial experimental design), time constraints and your greatest understanding will result from varying only one thing at a time. Thus, if you look at effect of pH, vary only pH and not the inoculum

Microbial Fuel Cells. By Bruce E. Logan

and container and temperature. Keep a laboratory notebook, taking careful notes especially when you change something in your experimental procedure that might affect your results later on. Take lots of pictures so you have a record of what something looked like in past experiments—and so you have some nice pictures for that science fair poster! Use safe laboratory practices, like wearing safety goggles and gloves. While it is unlikely that you will electrocute yourself (or even be able to feel the slight current generated by a homemade MFC), you will be growing bacteria and you never know what kind of bacteria will grow there (although we have no evidence of pathogens in our MFCs). Even if you wear gloves, be sure to wash your hands and other materials. Treat your MFC reactor waste like you would a baby's diaper—best to be moved quickly outside and to a suitable waste container. Also note that some science fairs have strict rules when it comes to handling bacteria. For example, the INTEL International Science and Engineering fair they have strict rules for all "projects involving microorganisms, recombinant DNA technologies and human or animal fresh tissues, blood or body fluids." They have two allowed classes of treatment, BSL-1 and BSL-2, with two higher classes of research that are prohibited due to safety concerns. They consider "the culturing of human or animal waste, including sewage sludge" as a BSL-2 study, requiring their strictest safety protocols. There are requirements for safe disposal of wastes, which includes autoclaving

A

A8 WEDNESDAY, MARCH 16, 2005 — 2005 Science & Engineering Fair

B

Sophomore takes top prize

Suzette Wenger / Intelligencer Journal photos

Abbie Groff, grand champion of the 2005 contest, is a sophomore at Conestoga Valley High School. She competed in the microbiology category.

Groff wins with microbiology project

BY REBECCA J. RITZEL
Intelligencer Journal Staff

If it weren't for Conestoga Valley sophomore Abbie Groff, the results of the 52nd annual Lancaster Newspapers Science & Engineering Fair would have been identical to the 51st annual fair.

Abbie remembers attending the science fair ceremony last year and watching Katie Nicholas, then a sophomore at Ephrata High School, win the grand champion trophy.

"I saw (Katie) win last year," Abbie said. "I remember thinking, 'Man, I want to win that.' I thought, maybe by senior year, my project will be good enough to win second place."

Entrants in Lancaster Newspapers Science & Engineering Fair who earned category and auxiliary awards will be listed in Thursday's edition of the Intelligencer Journal.

Tuesday night, Abbie was ecstatic to learn that, as a sophomore, she entered a project good enough to win the grand champion award—"Identification of Benthic Microbes Utilizing Bioremediation of Microbial Fuel Cells." She became only the third sophomore to carry home the tallest science fair trophy in the past 30 years.

GRAND CHAMPION

Eric Sauder, a Hempfield High School senior, was reserve champion this year and senior champion in 2004.

Katie and Abbie greeted each other enthusiastically Tuesday night after the awards ceremony. Both girls will travel with their projects to the 55th annual Intel International Science and Engineering Fair in Phoenix, Ariz.

Dressed in a flattering green suit, Abbie hardly fit the image of the stereotypical science geek, but she didn't deny that the microbiology project, with certain elements of physics, has been a major part of her life in the past two years.

"Oh my God, I am such a nerd," she said. "I am so excited."

The project is the second phase of a three-year study. Last year, she began studying the organic material and pollutants that congeal into sludge in local streams.

This year, she proved that electricity is produced as the material decomposes. During the next 12 months, she plans to search for a way to measure and channel that electricity.

Abbie said her parents, Herb and Deb Groff of Lancaster, are not "science people."

"I ended up teaching a lot of this to myself," she said. She also is consulting with professors at University of Massachusetts Amherst and Penn State University.

"By next year, this should be a really good project," Abbie said.

The 2005 Lancaster Newspapers Science & Engineering Fair was co-sponsored by Pfizer, Inc. and Pfizer Global Manufacturing,

Fig. 11.1 (A) Sediment MFC designed by Abbie Groff, and (B) newspaper clipping noting her winning the Lancaster County science fair. (Reprinted with permission from Lancaster Newspapers, Inc. March 16, 2005 issue of Intelligencer Journal.)

at 121°C for 20 minutes, disinfection with 10% sodium hypochlorite or alkaline hydrolysis, incineration, or biosafety pick-up. For more information, you can go to their website at www.sciserv.org/isef/rules/rules11.asp. Be sure to read these rules, as you would likely be disqualified from competition for not following them!

Success stories. Students from around the world have been experimenting with MFCs, ranging from simple sediment MFCs to more complex one- and two-chamber systems. One of the first students I had the pleasure to advise was Abbie Groff, a sophomore in high school in Lancaster County, PA. Abbie designed a sediment MFC using simple containers inoculated with mud from a nearby stream (Fig. 11.1). She was rewarded for her efforts with first-place, grand champion prize at the Lancaster County Science Fair. She went on to the INTEL competition that year in Phoenix Arizona, and got to enjoy seeing other young inventors and scientists. Sikandar Porter-Gill of Gaithersburg High School designed a two-chamber MFC using recycled plastic bottles and membranes and electrodes donated by different companies (Fig. 11.2). He placed first in the Environmental Science topic in the Montgomery County Area Science Fair, and received an ISEF award to go to the INTEL Fair in Indianapolis where he won a second-place award from the US Air Force. Many other students around the world have designed and tested MFCs. While we don't know the outcome of all those experiments, each time these experiments seem to get better and be more creative, and we look forward to seeing what new ideas will emerge from those efforts.

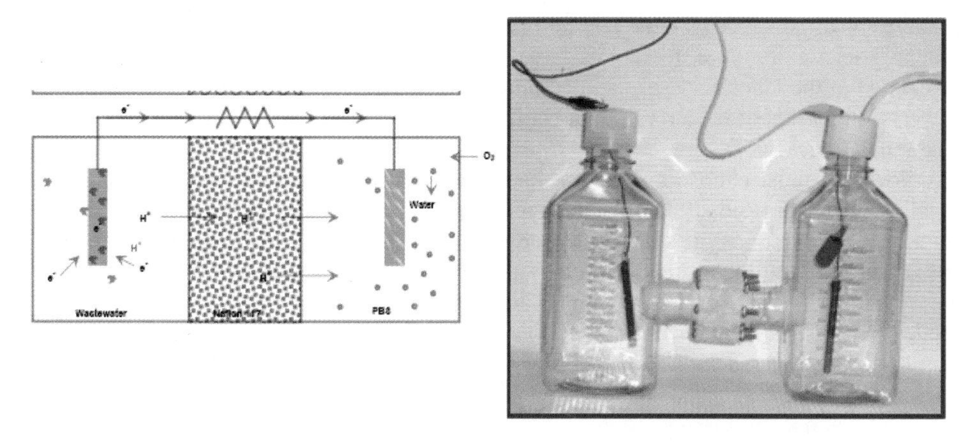

Fig. 11.2 (A) Schematic and (B) photograph of a two-chamber MFC designed by Sikandar Porter-Gill for the Montgomery County science fair. The purpose of Sikandar's project was to look at different membranes between the chambers.

The sections below were written with the idea guiding new researchers and inventors through the process of thinking about what type of MFC to design and build, and what questions might be interesting to explore. I have received requests from people around the world on how they could make an MFC with limited budgets or access to electrode materials, and thus I have tried to focus here on simple and inexpensive approaches. My goal is to stimulate your imagination about how to be creative yet successful on a small budget.

11.2 Choosing your inoculum and media

One of the first decisions you will need to make is where and how to obtain your starting culture, or bacterial inoculum. The source of bacteria for tests in my laboratory is a wastewater treatment plant because it contains such a diverse and plentiful supply of bacteria. You can view a wastewater treatment bioreactor as a type of engineered bacterial farm, with the treatment plant operators functioning as bacterial farmers, turning the waste they receive into bacteria (their crops!) and then transforming their crops into different final products (such as biosolids, used for land application and sometimes methane gas for power generation). However, science fair rules may prohibit using wastewater (see cautions above for sample origin and read the safety tips) so it is suggested that you find a more acceptable source of bacteria. A stream, river, or lake sediment is an excellent source of bacteria, and often the mud is anaerobic making it ideal for using in an MFC. Try to completely fill your container when getting a sample (no air space) and put it on ice to preserve the bacteria as much as possible in their current state. Use a plastic container as gas production by the sample could produce pressure that you don't want to have in a glass bottle. Other sources of bacteria should work fine, as it appears exoelectrogenic bacteria are well distributed in the environment. You might consider obtaining samples from several different sources, seeing which one works better in your tests. Are there food industries with wastes nearby that are typically used as animal feed? What about soils from a garden or your lawn. Even river and lake water contain about 10^5–10^6 bacteria per milliliter of water, and so you'll have no trouble finding bacteria for your tests. We have found that reactors that aren't intentionally inoculated in the laboratory, if operated under non-sterile conditions, eventually produce power. Getting lots of bacteria in your starting inoculum should just help your system get going more quickly.

Now you must choose a medium. First, consider your source of water. Tap water usually contains a residual disinfectant such as chlorine, so aerate it for at least a day to get the chlorine out (so you don't kill your bacteria). The water may be hard (*i.e.,* contain lots of calcium and magnesium) and may precipitate out needed metals when you make your media. That won't happen with distilled (DI) water, but all the trace metals often needed by bacteria have been removed to extremely low concentration in DI water, so it may not be optimal as well. You can always mix DI and tap water to save on DI water use.

Next, choose the ingredients for the medium. A microbiologist will often use a rich medium that ensures their culture will grow, and so many of the different media used by researchers for their studies contain far more chemicals than is absolutely necessary. A complex medium is one that has a composition that is not exactly known, such as Luria broth. With a defined medium, you know exactly what the medium will contain as you add only specific chemicals. Most standard rich media should work in MFC tests, so the real decision is what to use for a defined medium. In many tests we run we use a modified medium originally developed for iron-reducing bacteria. It contains quite a list of trace minerals and vitamins, but those are not known to all be needed. One simple medium that keeps well was used in early tests for power production in a two-chamber system based on modifying the composition of media used in biochemical oxygen demand (BOD) tests (*Logan et al.* 2005). This mineral salts medium (MSM) contained (mg/L in distilled water): NaCl, 8800; $MgCl_2{\cdot}7H_2O$, 330; $CaCl_2$, 275; KH_2PO_4, 14; K_2HPO_4, 21; $Na_2HPO_4{\cdot}7H_2O$, 56; $FeSO_4{\cdot}7H_2O$, 10; $MnSO_4{\cdot}H_2O$, 5; NH_4Cl, 3.1; KCl, 2;

$CoCl_2{\bullet}6H_2O$, 1; $ZnCl_2$, 1; $CuSO_4{\bullet}5H_2O$, 0.1; H_3BO_3, 0.1; Na_2MoO_4, 0.25; $NiCl_2{\bullet}6H_2O$, 0.24; EDTA, 1. The MSM also contained 3 g/L of $NaHCO_3$, but much less (say 0.1 g/L) can be used and too much bicarbonate in the medium will stimulate methanogen growth (they need high concentrations of carbon dioxide) and thus can reduce your final Coulombic efficiency (recovery of electrons as power). The solution pH was ~7. The substrate used should not result in a pH change of more than one unit from the start and during your experiment. The MSM can be prepared without the substrate at a 10-50× strength and diluted with water, and then added into your reactor when used. You can keep the MFC stock in a fridge, and it does not need to be kept sterile, but stir well before using as some of the metals may form a precipitate.

The ionic strength (IS) of the medium will affect solution conductivity, and IS has been shown to affect power generation (*Liu et al.* 2005a). The IS of this solution noted above is 164 mM, but the IS can be increased by adding more phosphate buffer or NaCl into the solution. Too high an IS will kill bacteria, but a low IS will limit power production. An empirical correlation of IS and total dissolved solids (which you can measure, or get from your water provider) is IS (M) = $2.5{\times}10^{-5}$ TDS (mg/L) (*Snoeyink and Jenkins* 1980). The conductivity of water can be measured with a meter, and a rough estimate of the IS with conductivity (γ) is IS (M) = $1.6{\times}10^{-5}\ \gamma\ (\mu\Omega^{-1}\ cm^{-1})$ (*Snoeyink and Jenkins* 1980). Note that conductivity units are often given as Siemens/cm (S/cm), where $S = 1/\Omega$.

If you have a low concentration of bacteria, you may want to add cysteine (L-cysteine HCl) as this will chemically scavenge dissolved oxygen so that the solution can be anoxic. Cysteine reacts with oxygen to form a disulfide dimer (cystine), according to

$$4\ C_3H_7NO_2S\text{-}H + O_2 \rightarrow 2\ C_3H_7NO_2S\text{-}SO_2NH_7C_3\ + 2\ H_2O \qquad (1)$$

Be careful not to add too much cysteine! Cysteine is an amino acid, and thus bacteria can use it as a sole substrate for growth rather than your intended substrate. For example, it was found that electricity generation was sustained in a two-chamber MFC at a cysteine concentration of 385 mg/L(0.5 g/L as L-cysteine HCl). Using the equation above, you can determine how much cysteine could be used to scavenge just the dissolved oxygen (7–9 mg/L) in solution, but remember oxygen will diffuse through the membrane between the chambers (or through the cathode in a single chamber system). Adding 50 mg/L would not be unreasonable at the beginning of an experiment, and it is not needed once power generation is shown to be sustained and at a good level for your device.

11.3 MFC materials: electrodes and membranes

The next challenge is choosing a material for the electrodes. Carbon paper and cloth is available from commercial manufacturers (*e.g.*, E-TEK) and easily obtained over the internet, but it can be quite expensive even for small pieces. When it is prepared with platinum for use as a cathode, it is even more expensive and thus likely to have a cost out of reach for most projects. Remember that any material that is conductive will work—you'd just prefer it didn't dissolve (corrode) away during your experiment. First, consider the material for the anode. Metals such as stainless steel will work, but bacteria don't seem to stick well to them at least in terms of generating power. Think about getting materials made of carbon or graphite. I once heard of a student that took apart 100 pencils and used the graphite core (the pencil lead) for his experiments. You can buy a bucket of conductive graphite at relatively low cost ($1/kg, *e.g.,* Graphite Sales, Inc., Chagrin Falls, OH), but be careful as the shipping is expensive (it is dense) and often minimum orders are needed (for example, 90 kg minimum). Graphite fibers cost about $40 for a single tow (50 yards long) which contain ample surface area for bacteria and it is lightweight (80,000 filaments, each only ~10 μm in diameter). Several vendors sell small amounts of graphite cloths and fibers (*e.g.*, Fibre Glast, OH). You can find carbon powders and coat any material to make it conductive—be creative on your mixes and choices of materials!

For the cathode, unless you can mix in your own Pt with a binder such as Nafion and carbon powder (*Cheng et al.* 2006c), you probably will have to do without a catalyst (*Logan et al.* 2005). Plus, you may need special ovens or conditions difficult to use catalysts other than Pt. You can still produce power without a catalyst, but it will be less than with a catalyst. Also, you can use a lot more surface area for the cathode than the anode, and thus when you normalize power to anode surface area or reactor volume, you may still get good results (*Oh and Logan* 2006). Trying different cathode materials or varying the properties and conditions of the cathode can be the great basis for an experiment.

Membranes. If you want to use a membrane for your MFC, you may need to be creative in choosing the material. Remember the membrane must be permeable to protons so a solid sheet of material (like aluminum foil) won't work. On the other hand, it can't be so porous that water will run right thorough it. In laboratory tests, we often use an ion exchange membrane like Nafion, but it costs far too much for a laboratory experiment. Less expensive ultrafiltration membranes will work as well but these do leak substrate between chambers (*Kim et al.* 2007b). A more readily available material is GoreTex™, which can be purchased at some fabric stores. This material will hold water and allow proton exchange. You might even have an old piece of clothing lying around that you could scavenge some GoreTex from for use in your reactor.

11.4 MFC architectures that are easy to build

Creating an MFC out of materials you may have available to you at home or at local stores is really the fun part of making the system. The first thing you need to decide is whether you want to make a single- or two-chamber system. A single-chamber system will probably produce more power, but it may be more difficult to control leakage of the reactor contents and there will be substrate lost to aerobic degradation of substrate by bacteria using oxygen that readily diffuses through the cathode. A two-chamber system will probably produce less power, but it is very easy to work with.

Single-chamber MFCs. You can make a single chamber MFC just by putting a cathode at the open end of a tube, with the anode inside the tube. For example, you can take a plastic bottle and cut out the lid leaving just the ring and a small part of the top that sits on the jar top to form a seal. A plastic Mason jar lid is perfect for this if you can find one as they already have a hole in the top (just discard the lid plate). Place your carbon cloth cathode over the top of the jar and screw the lid on, so that the lid holds the cloth to the end of the jar. You may need some extra sealant to keep water from running out. If you use a membrane, allow a bit of water between the membrane and cloth. Next, place the jar on its side and drill a hole in the top. Now you can insert your chosen anode material into the jar, and you can use this hole to fill/refill your reactor. Put a rubber stopper in it, or an S-shaped water seal device (available in beer or wine making supply stores) so that no oxygen can get into the chamber when you make it and carbon dioxide

Fig. 11.3 Single-chamber MFCs constructed by Eric Zielke at Humboldt State University using simple polycarbonate tubing and bolts.

can be easily released (especially important in large reactors). If you have access to a shop and some acrylic tubes and plates, then your design can be more sophisticated. For example, the single-chamber MFCs made by Eric Zielke at Humboldt State University used a polycarbonate tube with plates clamped onto either side (Fig. 11.3). This construction involved relatively little machining of the materials, but produced a simple and effective single-chamber system. Your local hardware store probably has PVC pipes which can serve the same purpose (although they are not transparent) as the polycarbonate tubing.

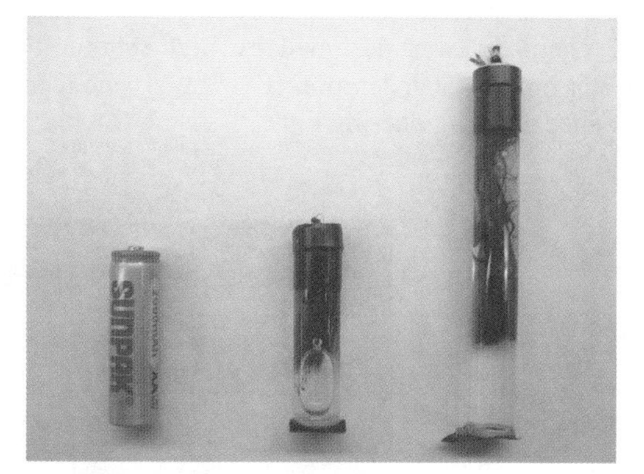

Fig. 11.4 Very simple single-chamber MFCs constructed by Garret Estadt of Penn State using a tube for the MFC housing, graphite fibers for the anode, and a carbon cloth cathode.

A simpler approach for the design of a single-chamber system was taken by Garett Estadt, an undergraduate student at Penn State. He used random graphite fibers for the anode, letting them disperse in a tube either 5 cm or 10 cm long (Fig. 11.4). The smaller tube reactor produced 86 mW/m^2 (1.73 W/m^3), while the 10-cm long reactor produced 33

mW/m^2 and 660 mW/m^3). The systems had high internal resistance (3700 Ω and 7000 Ω) limiting power production.

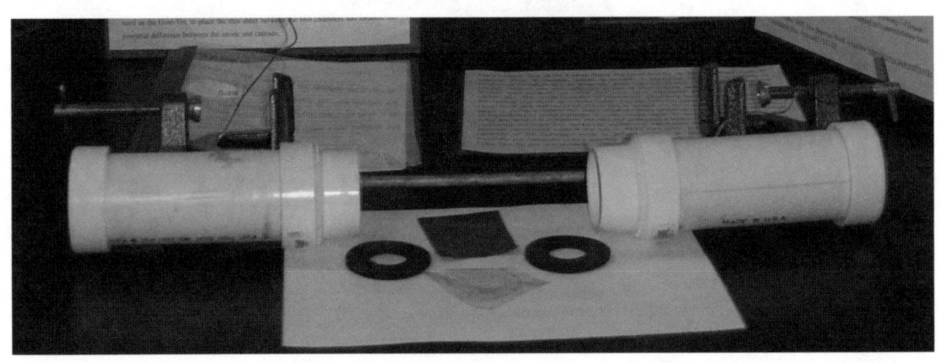

Fig. 11.5 Two-chamber MFC constructed by Ian Eschelman showing the electrode, membrane and gaskets used to clamp together the two tubes (the clamp is shown behind the tubes). Ian placed first in his science fair with this device.

Two-chamber MFCs. A two-chamber system may be easier to make and it will allow you to test a wider variety of materials for anodes, cathodes, and membranes. A couple of examples of two-chambered systems were shown in the photographs above, but many more ideas have been tried with this system. The key for this system is placing a membrane between the liquid contents of two chambers. You can use a tube to do this, as shown in the above examples, or you can just clamp the two tubes together. For example, a two chamber system was constructed by Ian Eschelman using large-diameter PVC pipes (Fig. 11.5). The membrane was held by two gaskets between the pipes, with a large clamp holding the two chambers together. A two-chamber system with a tube (centrifuge tubes) connecting the chambers was designed by high-school student Tim Chang to examine the effect of pH on power generation (Fig. 11.6). His results were presented at the New York City Science and Engineering Fair.

Fig. 11.6 Two-chamber systems designed by Tim Chang to examine the effect of pH on power generation. Note the pH meter is shown in the front, with the voltmeter in the background.

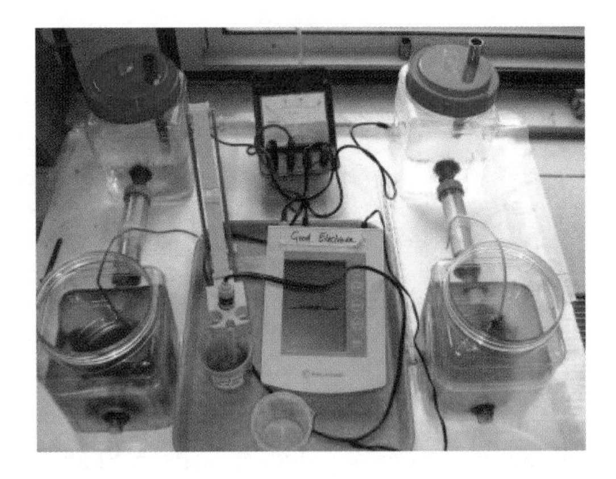

Fig. 11.7 A two-bottle flow through system based on using graphite granules as the anode, and carbon paper cathodes. The mesh (A) is clamped between the tubes, holding the graphite granules (B) in the anode chamber. (C) The reactor shown with the arrows indicating the flow path through the two chambers. The cathode chamber on the right is aerated with a diffuser.

Flow through systems. Various systems have been designed so that there is flow through the anode material. Jang *et al.* (2004) designed a simple system with flow through graphite felt anode towards the cathode, with no separator (other than glass wool) between the two chambers. A design based on this approach developed at Penn State (unpublished) is shown in Fig. 11.7. Here graphite granules are used as the anode, and a plastic mesh (like that used on a screen door) is used to hold the granules in the anode chamber. The cathode was carbon paper containing a Pt catalyst, although it is conceivable that other materials could be used here as well. The main problem with this design is that if the substrate is not all consumed in the anode, it goes into the cathode where bacteria grow aerobically. As shown in Fig. 11.7c, this resulted in the cloudy appearance of the bottle now covered with a biofilm.

11.5 MEC reactors

Instead of choosing to produce electricity, you may want to examine hydrogen production with an MEC or MEC reactor. All of the above information on MFCs applies to MECs, except that you will need to have a system that does not use oxygen anywhere, and one that allows for capture of the hydrogen gas produced in the cathode chamber. Hydrogen gas leaks easily through many materials, and couplings must be made tight to avoid gas leakage. You can make an MEC reactor with a two chamber MFC (*Liu et al.* 2005c), but you would need to flush both chambers with an inert gas to remove oxygen. Helium can be used for this, and you can buy containers of it from a local store specializing in parties and special events.

You will need to add potential to the circuit, and monitor current by placing a low-ohm (10 Ω) resistor in series with the power source and measuring the voltage across the resistor. You can buy a good power source from an electronics store, but an alternative and inexpensive way to apply voltage to an MEC circuit is using a potentiometer (Stephen Grot, Ion Power, Inc., *pers.. commun.*). A potentiometer is a variable resistor that is used to control the voltage output from a source of power, such as a battery (Fig. 11.8). Using a potentiometer, 0.5 volts can be applied to an MEC reactor using a 1.5-volt battery by simply adjusting the resistance control. Large-sized batteries (size D) are recommended because they will produce power for a longer time period than smaller ones (size AA). One disadvantage to using a potentiometer is that the voltage output decreases over time, and thus you will need to constantly monitor and adjust the voltage to ensure a near-constant voltage.

Fig. 11.8 A potentiometer, shown hooked up to a D-size battery, can be used to add voltage to the circuit in a MEC reactor. The voltage decreases over time, however, and must be constantly adjusted to maintain a near-constant applied voltage. [Photograph by D. Call.]

You can direct the gas produced from the cathode chamber through a tube and into an inverted buret with the open end placed into a beaker of water. As gas moves into the buret, it will displace water, allowing you to measure the volume of gas produced. You can run the gas first through a strong base solution to strip out CO_2, but there is not an easy way to remove methane if it is produced in the cathode. Unfortunately, there is no easy way to measure the composition of the gas for H_2, unless you have access to a gas chromatograph. Of course, be careful with hydrogen, as you could produce a composition of gas that is explosive! Work in a well-ventilated room, where releasing the gas will result in a low gas concentration of hydrogen that is not dangerous (it is not explosive below 4%).

11.6 Operation and assessment of MFCs

Once you've constructed your MFC, it is time to see how well it will perform. Inoculate the reactor with your source of bacteria and your medium, and then seal the anode chamber to prevent oxygen leaking in. Place a resistor in the circuit between the electrodes as the "load" for your MFC power supply (that is, don't just connect the two electrodes with a wire). You probably can't use an actual device, such as a light bulb or small motor (like a fan demonstrated on our website at www.engr.psu.edu/mfccam) unless you build a pretty big reactor. Also, the voltage will be low, not allowing you to light up a small LED type of light (which uses little power) unless you hook a couple of MFCs in series to bump up the voltage. To get the system going, try using a resistor in the range of 500 to 1000 Ω as this seems to work well in our tests. A low resistor (1–10 Ω) would intuitively seem to make sense as that should allow more current to flow, but we find this doesn't seem to be the case. (Maybe you would like to investigate this yourself.) Once the system is going, you can determine the maximum power by creating a polarization curve using a series of resistors, as described in Chapter 4 and elsewhere (*Logan et al.* 2006).

You can use a simple volt meter to measure the voltage drop across the resistor, allowing you to calculate current ($I = V/R$) and then power ($P = IV$). We use a multimeter, which allows us to measure the voltage at set time intervals (20 to 30 minutes) from 20 to 40 different systems, but these meters are expensive and require a computer for data storage. On the other hand, if you have access to a computer with any type of data acquisition system you could probably use that for your MFC. A data acquisition system that is hooked up to a device in the laboratory usually just measures the signal in terms of a voltage (an analog input), usually ranging from 0 to 1 V or to 10 V. A simple analog-to-digital conversion board can be obtained from many electronic supply shops (to store the value as a number), allowing you to create your own data acquisition system as a part of your experimental apparatus.

Remember, the key to any scientific investigation is a good hypothesis to test. So be sure to have a testable outcome. For example, a hypothesis like "a pH in a slightly acid region (pH = 5.5–6.5) will improve power generation by increasing the proton concentration in solution" as opposed to "the effect of pH on power generation over the pH range of 5 to 8". The same experiments can be conducted, but you have a specific hypothesis in the first approach. Remember to test the open-circuit voltage, or the voltage between the electrodes when the circuit is disconnected. This will be your maximum possible voltage, and it represents the voltage with an infinite resistance. Your highest power will occur when your external resistor is equal to the internal resistance of the system (see Chapter 4 for more information on this). In general, if your voltage is high keep decreasing the resistor and computing power to identify the maximum power. This would be a good resistance to use when you run the system for your tests. Producing as much power as you can is probably the single goal of any MFC design and test. But the most important thing to remember is to *Have Fun*!

CHAPTER 12

Outlook

12.1 MFCs yesterday and today

While the concept of bioelectricity generation was first demonstrated nearly a century ago, MFCs as we now know them from recent work really need to be considered as a new technology. Biofuel cells conducted with yeast and bacteria that needed chemical mediators to be added to the reactor were very unlikely to have practical applications. Thus, modern MFCs can be considered to have only emerged in 1999 with the finding of electricity generation without the need for exogenous mediators (*Kim et al.* 1999c). Since then, substantial progress has been made in the field despite the fact that only a relatively few research laboratories have been working in this area. We can expect that as more researchers engage in improving MFC technologies that advances will be made even more quickly.

In the past 8 years, we have seen a nearly logarithmic increase in power production with MFCs using oxygen at the cathode (Fig. 12.1). Before 2001, power production was less than 0.1 W/m^2 of electrode. In 2001 several reports emerged on systems with power over 10 mW/m^2 (normalized to anode surface area), and in recent years we have seen power levels over 1000 mW/m^2, with a maximum value achieved of 2400 mW/m^2 (normalized to the cathode surface area) (*Logan et al.* 2007). This is still at least an order-of-magnitude below power densities that could theoretically be achieved before mass transfer to the biofilm would limit power densities.

Our attention is now shifting from power normalized to electrode surface area, to power per volume. This transition reflects a growing appreciation by engineers that this technology is ready to emerge into practical applications. Most early MFC studies were not conducted to optimize volumetric power density but rather to explore materials and understand factors that limited power on an area basis. As systems have been designed more with a desire to improve performance on the basis of reactor volume, power volumetric power densities have increased for MFCs using oxygen at the cathode, from values below 1 W/m^3 only a few years ago to 115 W/m^3 today (*Cheng and Logan* 2007; *Logan and Regan* 2006a). We can expect these power densities to continue to increase in the next few years.

It has been suggested that a goal for MFCs is to reach a power volumetric density comparable to that produced with anaerobic digesters (*Rabaey and Verstraete* 2005), or ~400 W/m^3. While that is a reasonable goal, hydraulic retention times, energy used for pumping, and materials costs all we need to be included in this analysis. A goal for MFCs described in Chapter 9 is to achieve HRT comparable to that used for an activated sludge reactor, or an HRT of about 4–6 hours. Anaerobic digesters used for sludge treatment can have detention times of a month or more, and thus are impractical for treatment of dilute streams such as those typical of domestic wastewater treatment plants and many industrial wastewaters. The real goal of an MFC design will be one that is economical on the basis of capital and operational costs compared to that needed for conventional designs.

12.2 Challenges for bringing MFCs to commercialization

There are many challenges remaining to fully exploit the maximum power production possible by MFCs, to find ways to make the systems economical, and to create wastewater treatment systems based on MFC bioreactors. Power densities still need to be increased but this must be done under realistic conditions. For example, work with chemical catholytes such as ferricyanide should be abandoned and the focus should be squarely placed on using oxygen in air at the cathode. Materials, and different methods to treat materials, must be examined that are efficient both in terms of power generation and cost. It is clear that anode treatment can reduce acclimation times and increase power

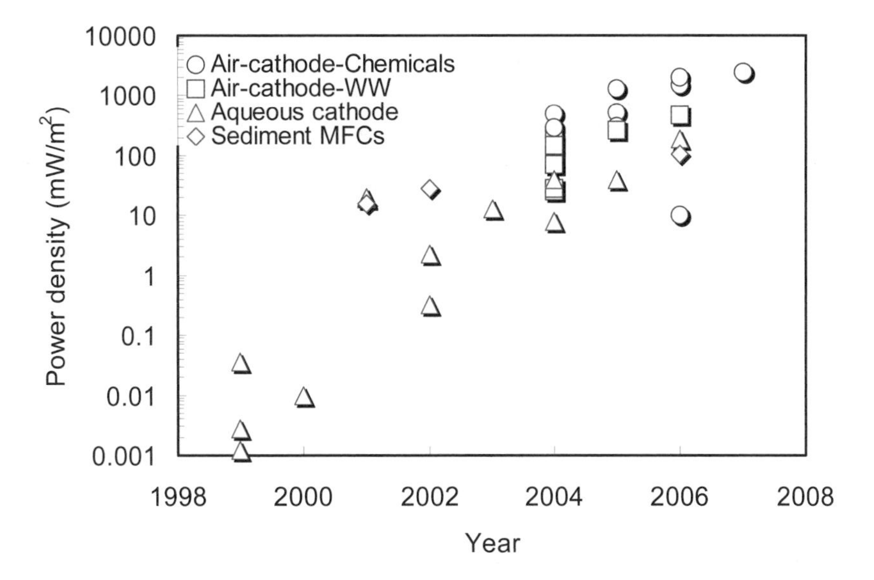

Fig. 12.1 Power production reported since 1999 for MFCs using oxygen at the cathode. Not included are biohydrogen MFCs or reactors with sacrificial anodes, or data from systems using chemical catholytes such as ferricyanide or permanganate. [Adapted from *Logan and Regan* (2006a).]

densities, but further improvements are needed in simplicity and costs of anode pre-treatment. The greatest challenge is for cost-effective cathodes, as it currently appears that this component will ultimately control the overall cost of the system. Tubular cathodes appear to be a viable architecture solution, but an optimum material and conductive coating are still needed. There are still many different arrangements and reactor designs possible even with brush anodes and tube cathodes, and much work is needed in optimizing system architecture.

Despite the potential of the MFC as a method of wastewater treatment, most researchers have chosen to work with defined compounds such as glucose or acetate. Using these compounds is necessary to understand the response of the system uncomplicated by the variations in wastewater. However, most of the organic matter in wastewater is protein, and the bulk of the material is poorly defined and will likely remain so. Much more work is needed on actual wastewaters in order to better understand how designs will function when given the complex and particle laden waters. We need to focus on reactor factors that govern the cost of treatment systems, such as hydraulic and solids retention times (HRTs and SRTs), the mass of biomass produced and its settleability or filterability in engineered systems, and the impact of recycle streams on system performance. More information is needed on the flow of nutrients, such as nitrogen and phosphorous in the system, and methods to control these in MFC-based systems. Pilot reactors need to be designed and tested so that we can be more realistic data on system performance. The potential for system failure due to biofouling or loss of conductivity of the medium need to be explored with materials considered for use in the system. As the MFC is inherently an anaerobic process, we need additional information on how to control the community to shift it towards electrogenesis, and away from sulfate reduction and methanogenesis.

In industrial applications, it may be possible to better control the types of bacteria in a process or even use pure cultures. In all systems, we need better information on the microorganisms that can function as exoelectrogens. What are the most useful strains, and which bacteria can produce the most power for us? How can we manipulate these strains for efficient cellulose degradation and maximize hydrogen recovery in an MEC or BEAMR process? How thick can a biofilm become before it is no longer useful as a method of electricity generation, and what signals trigger nanowire production in cells? Is there a natural evolution of bacteria in an electrogenic biofilm, and, if so, what controls that and what bacteria emerge as the "winners"? Clearly, there is much exciting work to be done on better understanding the bacteria that function within an electrogenic biofilm and that help us to harvest electricity from MFCs.

12.3 Accomplishments and outlook

The list of accomplishments in our understanding of how electricity is produced in an MFC, and how costs can be reduced while power densities are increased, are impressive. We have seen designs based on aqueous cathodes, where oxygen must be dissolved in water (*Bond et al.* 2002; *Bond and Lovley* 2003), give way to air cathodes which produce more power and are less expensive to operate as oxygen transfer is passive (*Liu et al.* 2005c; *Liu and Logan* 2004; *Liu et al.* 2004). Precious metals such as platinum are no longer needed on either electrode, with non-precious metals such as cobalt and iron producing comparable power densities when used as a catalyst for oxygen reduction (*Cheng et al.* 2006c; *Logan et al.* 2007; *Zhao et al.* 2005). Expensive electrode materials such as carbon cloth, carbon paper, and graphite rods (*Bond and Lovley* 2003; *Rabaey et al.* 2003) have been replaced with less expensive ones, and higher surface area materials such as graphite fibers and graphite granules (*Logan et al.* 2007; *Rabaey et al.* 2005b; *Rabaey et al.* 2006). Expensive membranes such as Nafion, have been replaced with relatively inexpensive ion exchange membranes based on either cation or anion chemical species (*Kim et al.* 2007b) or even completely eliminated (*Liu and Logan* 2004). Cathodes have been improved by application of diffusion layers that control oxygen intrusion into the anode system and help avoid water losses, while at the same time improving power densities (*Cheng et al.* 2006a). Anodes have been shown to increase in performance by addition of organic and inorganic materials such as mediators and modifications such as a high-temperature ammonia-treatment processes (*Cheng and Logan* 2007; *Lowy et al.* 2006; *Park and Zeikus* 2003). And finally, a plan for a scalable system based on graphite brush anodes and tubular cathodes has opened the door to a reactor configuration that should become efficient in terms of power production, easily scaled, and able to be operated in a manner needed for wastewater treatment (*Logan* 2005; *Logan et al.* 2007; *Zuo et al.* 2007).

MFC applications are just beginning in areas other than wastewater treatment. Modified MFCs, called MECs or BEAMRs, can be used to achieve biohydrogen production from virtually any biodegradable organic matter. Recent advances have produced acetate to hydrogen conversion efficiencies of 82% from acetate, a compound once considered to be a "dead end" for biological hydrogen production. With glucose, this means shattering the "fermentation barrier," moving from a maximum possible conversion of 4 mol-H_2/mol-glucose, and typical conversion efficiencies of 2 mol/mol or less, to 10 mol/mol conversion efficiencies or more. The generation of power in remote location, once limited by battery lifetimes, can now be extended with sediment MFCs running off the organic matter in the sediment, or particulate organic matter included in the reactor (*Rezaei et al.* 2007). Bioremediation may also be possible with the very same MFC-based technologies, with applications possible for both chemical oxidation at the anode (organic pollutant oxidation) and chemical reduction at the cathode (inorganic chemicals such as nitrate and uranium) (*Gregory and Lovley* 2005).

The ultimate achievement in for MFCs will be when they can be used solely as a method of renewable energy production. Right now, the high costs of materials for MFCs and the relatively cheap price of fossil fuels makes it unlikely that electricity production can be competitive with existing energy production methods. However, MFCs are carbon neutral and power can be generated with cellulosic materials (*Ren et al.* 2007; *Rismani-Yazdi et al.* 2007). Thus, advancements in power densities, reductions in materials costs, and a global need to produce power without net CO_2 emissions may one day make MFCs practical just for electricity production. Much work remains to be done, but hopefully you agree that there is a bright and promising future for a wide range of MFC technologies that can form the foundation of a new generation of electrogenic reactor systems.

Notation

Notation with typical units

A_i Area, with various subscripts: *An* = anode, *Cat* = cathode, *m* = membrane; *cell* = bacterial cell, projected (cm^2)

b_i A constant, with various subscripts: *eO* = electrons/mol-O_2; *es* = electrons per mole of substrate; H2/*s* = stoichiometric moles of H_2 per mole of substrate; *N* = NADH/NAD^+ ratio

B_i Dimensionless constant with various subscripts: 0 = zero-order reaction; 1 = first-order reaction ( )

c_i Concentration of a chemical, with various subscripts: *An* = chemical in the anode chamber; *b0* = at the biofilm surface; *Cat* = chemical in the anode chamber; *eff* = sBOD concentration in the effluent of a solids contact process; *g* = gas; H2,*An* = H_2 in the water in the anode; H2,*Cat* = H_2 in the water in the cathode; H2*g* = H_2 in the gas; *inf* = sBOD concentration in the influent of a solids contact reactor; *s* = substrate; (g/L)

c^* Concentration, dimensionless ( )

C_E Coulombic efficiency ( )

COD Chemical oxygen demand (g/L)

D_i Chemical diffusivity, with various subscripts: *Am* = acetate in a membrane; *Cb* = chemical *C* in a biofilm *b*; *Cm* = chemical *C* in a membrane *m*; *Cw* = chemical *C* in water *w*; H2*w* = hydrogen in water; H2*m* = hydrogen in membrane; *Om* = oxygen in a membrane (cm^2/s)

e_i Ratio of energy input in BEAMR to total energy of electricity and substrate, with various subscripts: *in* = input in electricity; *S* = input in substrate ( )

E Voltage (V)

E^0 Standard electromotive force (V)

$E^{0/}$ Electromotive force under specific conditions (V)

E_i Electromotive force, with various subscripts: *An* = anode; *Cat* = cathode; *MFC* = measured across the cell in an MFC; *emf* = maximum (V)

E_{ps} Applied voltage of a power source (V)

$E_i^{0/}$	Electromotive force under standard conditions for: *An* = anode, *Cat* = cathode (V)
f_{DW}	Constant, indicating fraction of dry weight of a bacterial cell ( )
F	Faraday's constant (96,485 C/mol)
G	Gibbs free energy (J)
G^0	Gibbs free energy under standard conditions (J)
h	Number of homes (#)
H	Enthalpy, where ΔH is the change in enthalpy (kJ/mol)
HRT	Hydraulic retention time (d)
H_{H2}	Henry's law constant for hydrogen ( )
I	Current (A)
I_i	Current, with various subscripts: *An* = normalized to the anode area (A/m^2); *Cat* = normalized to the cathode area (A/m^2); *v* = normalized to the reactor volume (A/m^3)
J_i	Chemical flux, with various subscripts: *b* = flux to a biofilm; *CA* = cathode to anode chamber; *m* = flux through a membrane (g/m^2-s)
k_0	Rate constant, zero order (L/g-s)
k_1	Rate constant, first order (1/s)
$\mathbf{k}_i$	Specific phase mass transfer coefficient, with various subscripts: *Cm* = unspecified chemical in the membrane; *Cw* = unspecified chemical in water; *Om* = oxygen in the membrane; *Am* = acetate in the membrane; *H2m* = hydrogen in the membrane; *H2w* = hydrogen in the water (cm/s)
$\mathbf{K}_{CA}$	Overall mass transfer coefficient for hydrogen in water, from the cathode to the anode chamber (cm/s).
K_c	Half-saturation constant for the Monod growth rate equation (g/L)
l_i	Length, with various subscripts: *An* = anode, *Cat* = cathode (cm)
m_i	Mass, with various subscripts: H2 = hydrogen; *cell* = bacterial cell (g)
M_i	Molecular weight of various chemical species: H2 = hydrogen; *s* = substrate (g/mol)
n	Number, as in number of electrons transferred (#)
n_i	Number of moles of some chemical species *i*: *CE* = moles of H_2 that could be produced from the measured current; H2 = hydrogen; *s* = substrate; *th* = theoretical number of moles of H2 produced from a given mass of substrate (#)
N_b	Number concentration of bacteria (#/ml)
OCP	Open-circuit potential (V)
OCV	Open-circuit voltage (V)
OCV^*	*OCV* as calculated from the linear portion of the polarization curve (not a true *OCV*) (V)
OP_i	Overpotential, with various subscripts: *An* = Anode, *Cat* = cathode (V)
p	pressure
P	Power (W)
P_{An}	Power, normalized to the area of the anode (W/m^2)
P_{Cat}	Power, normalized to the area of the cathode (W/m^2)
P_{max}	Power, total maximum power output (*i.e.,* useful power) based on the measured *OCV* (W)
$P_{max,Cat}$	Power, maximum power normalized to the cathode area (W/m^2)

$P_{max,emf}$ Power, total maximum power output (*i.e.,* useful power) based on the calculated E_{emf} (W)
$P_{t,emf}$ Power, total maximum based on the calculated E_{emf} (W)
$P_{t,OCV}$ Power, total maximum based on the measured *OCV* (W)
P_v Power, normalized by the reactor volume (W/m^3)
q Liquid flowrate (m^3/s)
Q_{H2} Hydrogen gas flowrate (m^3/s)
r_i Recovery of hydrogen, with various subscripts: *Cat* = cathodic recovery; *CE* = Coulombic recovery ($r_{CE} = C_E$); H2 = overall recovery (%)
R Resistance (Ω)
R_i Resistance, with various subscripts: *ext* = external resistance; *int* = internal resistance; Ω = ohmic resistance (Ω)
R Gas constant (8.31447 J/mol-K; 0.0821 L-atm/mol-K)
t Time (s)
t_i Time with various subscripts: *b* = batch cycle; *d* = doubling time for bacterial growth (s, h or d)
T Temperature (K)
v Volume (m^3)
v_i Volume with various subscripts: *An* = anode chamber; *Cat* = cathode chamber; *cell* = bacterial cell (m^3); *L* = liquid (L)
V_Ω Ohmic loss due to solution conductivity (V)
w_i Width, with various subscripts: *An* = anode, *Cat* = cathode (cm)
W_i Energy, with various subscripts: H2 = equivalent for moles of hydrogen produced; H2,*max* = equivalent for theoretical maximum moles of hydrogen produced; *In* = energy input after correcting for loss to external resistor; *ps* = power source energy input; *R* = loss to external resistor; *s* = energy equivalent of substrate consumed (kWh)
X Cell concentration, in various units of *DW* = dry weight, *VS* = volatile solids, or protein (mg/L)
Y_{H2} Yield of hydrogen (usually dimensionless, but other units possible, *e.g.,* ml-H_2/g-COD)
$Y_{X/c}$ Yield constant for cells: cell mass, *X*, produced per mass of chemical substrate, *c* ( )

Greek letters (typical units)

δ_i Thickness with various subscripts: *s* = stagnant film; *b* = biofilm; *w* = water (cm)
η_E Energy efficiencies, with various subscripts: *E* = MFC based on electricity produced; *W* = BEAMR based on electricity input; *s* = BEAMR based on substrate input; *W+s* = BEAMR based on electricity and substrate ( )
μ Monod growth rate constant (1/s, 1/h, 1/d)
μ_{max} Monod growth rate constant, maximum value (1/s, 1/hr, 1/d)
Π Products of reaction ( )
ρ_{cell} Density of a bacterial cell (g/cm^3)
σ Solution conductivity (S/cm)
φ Fraction of substrate not accounted for by Coulombic efficiency, C_E and cell yield, $Y_{X/c}$ ( )

References

Aelterman, P., Rabaey, K., Pham, T.H., Boon, N. and Verstraete, W. (2006) Continuous electricity generation at high voltages and currents using stacked microbial fuel cells. Environ. Sci. Technol. 40, 3388-3394.

Alberty, R.A. (2003) Thermodynamics of biochemical reactions, John Wiley & Sons, New York.

Allen, R.M. and Bennetto, H.P. (1993) Microbial fuel cells: electricity production from carbohydrates. Appl. Biochem. Biotechnol. 39(2), 27-40.

Atchison, J.E. and Hettenhaus, J.R. (2004) Innovative methods for corn stover collecting, handling, storing and transporting, NREL/SR-510-33893; National Renewable Energy Laboratory: Golden, CO, 2004.

Bard, A.J. and Faulkner, L.R. (2001) Electrochemical Methods: Fundamentals and Applications, John Wiley & Sons, New York.

Beliaev, A.S. and Saffarini, D.A. (1998) *Shewanella putrefaciens mtrB* encodes an outer membrane protein required for Fe(III) and Mn(IV) reduction. J. Bacteriol. 180(23), 6292-6297.

Beliaev, A.S., Saffarini, D.A., McLaughlin, J.L. and Hunnicutt, D. (2001) MtrC, an outer membrane decahaem c cytochrome required for metal reduction in *Shewanella putrefaciens* MR-1. Molec. Microbiol. 39(3), 722-730.

Bergel, A., Feron, D. and Mollica, A. (2005) Catalysis of oxygen reduction in PEM fuel cell by seawater biofilm. Electrochem. Commun. 7(9), 900-904.

Bond, D.R., Holmes, D.E., Tender, L.M. and Lovley, D.R. (2002) Electrode-reducing microorganisms that harvest energy from marine sediments. Science 295(5554), 483-485.

Bond, D.R. and Lovley, D.R. (2003) Electricity production by *Geobacter sulfurreducens* attached to electrodes. Appl. Environ. Microbiol. 69(3), 1548-1555.

Cai, M., Liu, J. and Wei, Y. (2004) Enhanced biohydrogen production from sewage sludge with alkaline pretreatment. Environ. Sci. Technol. 38(11), 3195-3202.

Chaudhuri, S.K. and Lovley, D.R. (2003) Electricity generation by direct oxidation of glucose in mediatorless microbial fuel cells. Nat. Biotechnol. 21(10), 1229-1232.

Cheng, S., Liu, H. and Logan, B.E. (2006a) Increased performance of single-chamber microbial fuel cells using an improved cathode structure. Electrochem. Commun. 8, 489-494.

Cheng, S., Liu, H. and Logan, B.E. (2006b) Increased power generation in a continuous flow MFC with advective flow through the porous anode and reduced electrode spacing. Environ. Sci. Technol. 40, 2426-2432.

Cheng, S., Liu, H. and Logan, B.E. (2006c) Power densities using different cathode catalysts (Pt and CoTMPP) and polymer binders (Nafion and PTFE) in single chamber microbial fuel cells. Environ. Sci. Technol. 40, 364-369.

Cheng, S. and Logan, B.E. (2007) Ammonia treatment of carbon cloth anodes to enhance power generation of microbial fuel cells. Electrochem. Commun. 9(3), 492-496.

Childers, S.E., Ciufo, S. and Lovley, D.R. (2002) *Geobacter metallireducens* accesses insouble Fe(III) oxide by chemotaxis. Nature 416, 767-769.

Choi, Y., Jung, E., Park, H.S., Paik, S.R. and Jung, S., Kim, S. (2004) Construction of microbial fuel cells using thermophilic microorganisms, *Bacillus licheniformis* and *Bacillus thermoglucosidasius*. Bull. Korean Chem. Soc. 25(6), 813-818.

Clauwaert, P., Rabaey, K., Aelterman, P., De Schamphelaire, L., Pham, T.H., Boeckx, P., Boon, N. and Verstraete, W. (2007) Biological denitrification in microbial fuel cells. Environ. Sci. Technol. 41(9), 3354-3360.

Cooper, K.R. and Smith, M. (2006) Electrical test methods for on-line fuel cell ohmic resistance measurement J. Power Sour. 160, 1088-1095.

Cooper, N.B., Marshall, J.W., Hunt, K. and Reidy, J.G. (2007) Less power, great performance. Wat. Environ. Technol. 19(2), 63-66.

Cord-Ruwisch, R., Lovley, D.R. and Schink, B. (1998) Growth of *Geobacter sulfurreducens* with acetate in syntrophic cooperation with hydrogen-oxidizing anaerobic partners. Appl. Environ. Microbiol. 64, 2232-2236.

Crittenden, S.R., Sund, C. and Sumner, J.J. (2006) Mediating electron transfer from bacteria to a gold electrode via a self-assembled monolayer. Langmuir 22, 9473-9476.

Datar, R., Huang, J., Maness, P.C., Mohagheghi, A., Czernik, S. and Chornet, E. (2006) Hydrogen production from the fermentation of corn stover biomass pretreated with a steam explosion process. Int. J. Hydrogen Energy 32(8), 932-939.

Ditzig, J., Liu, H. and Logan, B.E. (2007) Production of hydrogen from domestic wastewater using a bioelectrochemically assisted microbial reactor (BEAMR). Int. J. Hydrogen Energy 32(13), 2296-2304.

Dobbin, P.S., Powell, A.K., McEwan, A.G. and Richardson, D.J. (1995) The influence of chelating-agents upon the dissimilatory reduction of Fe(III) by *Shewanella putrefaciens*. Biometals 8(2), 163-173.

Doong, R.A. and Schink, B. (2002) Cysteine-mediated reductive dissolution of poorly crystalline iron(III) oxides by *Geobacter sulfurreducens*. . Environ. Sci. Technol. 36(2939-2945).

Emde, R., Swain, A. and Schink, B. (1989) Anaerobic oxidation of glycerol by *Escherichia coli* in an amperometric poised-potential culture system. Appl. Microbiol. Biotechnol. 32, 170-175.

Finkelstein, D.A., Tender, L.M. and Zeikus, J.G. (2006) Effect of electrode potential on electrode-reducing microbiota. Environ. Sci. Technol. 30(22), 6990-6995.

Freguia, S., Rabaey, K., Yuan, Z. and Keller, J. (2007) Electron and carbon balances in microbial fuel cells reveal temporary bacterial storage behavior during electricity generation Environ. Sci. Technol. 41(8), 2915-2921.

Girbal, L., Croux, C., Vasconcelos, I. and Soucaille, P. (1995) Regulation of metabolic shifts in *Clostridium acetobutylicum* ATCC 824. FEMS Microbiol. Rev. 17, 287-297.

Glassner, D.A., Hettenhaus, J.R. and Scheichinger, T.M. (1999) Perspectives on New Crops and New Uses Janik, J. (ed), pp. 74-82, ASHA Press, Alexandria, VA.

Gorby, Y.A. and Beveridge, T.J. (2005) Composition, reactivity, and regulation of extracellular metal-reducing structures (nanowires) produced by dissimilatory metal reducing bacteria, Warrenton, VA.

Gorby, Y.A., Yanina, S., McLean, J.S., Rosso, K.M., Moyles, D., Dohnalkova, A., Beveridge, T.J., Chang, I.S., Kim, B.H., Kim, K.S., Culley, D.E., Reed, S.B., Romine, M.F., Saffarini, D.A., Hill, E.A., Shi, L., Elias, D.A., Kennedy, D.W., Pinchuk, G., Watanabe, K., Ishii, S., Logan, B.E., Nealson, K.A. and Fredrickson, J.K. (2006) Electrically conductive bacterial nanowires produced by *Shewanella oneidensis* strain MR-1 and other microorganisms. PNAS 103(30), 11358-11363.

Gottschalk, G. (1986) Bacterial Metabolism, Springer-Verlag, New York.

Grant, P.M. (2003) Hydrogen lifts off—with a heavy load. Nature 424, 129-130.

Gregory, K.B., Bond, D.R. and Lovley, D.R. (2004) Graphite electrodes as electron donors for anaerobic repiration. Environ. Microbiol. 6, 596-604.

Gregory, K.B. and Lovley, D.R. (2005) Remediation and recovery of uranium from contaminated subsurface environments with electrodes. Environ. Sci. Technol. 39, 8943-8947.

Hasvold, Ø., Henriksen, H., Melvær, E., Citi, G., Johansen, B.Ø., Kjønigsen, T. and Galetti, R. (1997) Sea-water battery for subsea control systems. J. Power Sour. 65, 253-261.

Hasvold, Ø., Johansen, K.H., Mollestad, O., Forseth, S. and Størkersen, N. (1999) The alkaline aluminum/hydrogen peroxide power source in the Hugin II unmanned underwater vehicle. J. Power Sour. 80, 254-260.

Hawkes, F.R., Hussy, I., Kyazze, G., Dinsdale, R. and Hawkes, D.L. (2007) Continuous dark fermentative hydrogen production by mesophilic microflora: Principles and progress. Int. J. Hydrogen Energy 32, 172-184.

He, Z. and Angenent, L.T. (2006) Application of bacterial biocathodes in microbial fuel cells. Elecroanal. 18(19-20), 2009-2015.

He, Z., Minteer, S.D. and Angenent, L.T. (2005) Electricity generation from artificial wastewater using an upflow microbial fuel cell. Environ. Sci. Technol. 39(14), 5262-5267.

He, Z., Wagner, N., Minteer, S.D. and Angenent, L.T. (2006) The upflow microbial fuel cell with an interior cathode: assessment of the internal resistance by impedance spectroscopy. Environ. Sci. Technol. 40(17), 5212-5217.

Heidelberg, J.F., Paulsen, I.T., Nelson, K.E., Gaidos, E.J., Nelson, W.C., Read, T.D., Elisen, J.A., Seshadri, R., Ward, N., Methe, B., Clayton, R.A., Meyer, T., Tsapin, A., Scott, J., Beanan, M., Brinkac, L., Daugherty, S., DeBoy, R.T., Dodson, R.J., Durkin, A.S., Haft, D.H., Kolonay, J.F., Madupu, R., Peterson, J.D., Ymayam, L.A., White, O., Wolf, A.M., Vamathevan, J., Weidman, J., Impraim, M., Lee, K., Berry, K., Lee, C., Mueller, J., HKhouri, H., Gill, J., Utterback, T.R., McDonald, L.A., Feldblyum, T.V., Smith, H.O., Venter, J.C., Nealson, K.H. and Fraser, C.M. (2002) Genome sequence of the dissimilatory metal ion-reducing bacterium *Shewanella oneidensis*. Nat. Biotechnol. 20, 1118-1123.

Heilmann, J. and Logan, B.E. (2006) Production of electricity from proteins using a single chamber microbial fuel cell. Water Environ. Res. 78(5), 531-537.

Hernandez, M.E., Kappier, A. and Newman, D.K. (2004) Phenazines aqnd other redox-active antibiotics promote microbial mneral reduction Appl. Environ. Microbiol. 79(2), 921-928.

Holmes, D.E., Bond, D.R. and Lovley, D.R. (2004a) Electron transfer by *Desulfobulbus propionicus* to Fe(III) and graphite electrodes. Appl. Environ. Microbiol. 70(2), 1234-1237.

Holmes, D.E., Bond, D.R., O'Neil, R.A., Reimers, C.E., Tender, L.R. and Lovley, D.R. (2004b) Microbial communities associated with electrodes harvesting electricity from a variety of aquatic sediments. Microb. Ecol. 48(2), 178-190.

Holmes, D.E., Nicoll, J.S., Bond, D.R. and Lovley, D.R. (2004c) Potential role of a novel psychrotolerant member of the family *Geobacteraceae*, *Geopsychrobacter electrodiphilus* gen. nov., sp. nov., in electricity production by a marine sediment fuel cell. Appl. Environ. Microbiol. 70(10), 6023-6030.

Hurwitz, H.D. and Dibiani, R. (2001) Investigation of electrical properties of bipolar membranes at steady state with transient methods. Electrochim. Acta 47, 759-773.

Jang, J.K., Pham, T.H., Chang, I.S., Kang, K.H., Moon, H., Cho, K.S. and Kim, B.H. (2004) Construction and operation of a novel mediator- and membrane-less microbial fuel cell. Process Biochem. 39(8), 1007-1012.

Jiang, J. and Kucernak, A. (2004) Investigations of fuel cell reactions at the composite microelectrode|solid polymer electrolyte interface. I. Hydrogen oxidation at the nanostructured Pt|Nafion® membrane interface. J. Electroanal. Chem. 567, 123-137.

Jin, S. and Morris, J. (2007) The feasibility of using microbial fuel cell technology in bioremediation of hydrocarbons in groundwater. Environ. Sci. Health A., In press.

Jong, B.C., Kim, B.H., Chang, I.S., Liew, P.W.Y., Choo, Y.F. and Kang, G.S. (2006) Enrichment, performance and microbial diversity of a thermophilic mediatorless microbial fuel cell. Environ. Sci. Technol. 40(20), 6449-6454.

Kim, B.-H., Ikeda, T., Park, H.-S., Kim, H.-J., Hyun, M.-S., Kano, K., Takagi, K. and Tatsumi, H. (1999a) Electrochemical activity of an Fe(III)-reducing bacterium, *Shewanella putrefaciens* IR-1, in the presence of alternative electron acceptors. Biotechnol. Techniques 13(7), 475-478.

Kim, B.H., Kim, H.-J., Hyun, M.-S. and Park, D.-H. (1999b) Direct electrode reaction of Fe(III)-reducing bacterium, *Shewanella putrefaciens*. J. Microbiol. Biotechnol. 9(2), 127-131.

Kim, B.H., Park, D.H., Shin, P.K., Chang, I.S. and Kim, H.J. (1999c) Mediator-less biofuel cell. U.S. Patent 5976719.

Kim, B.H., Park, H.S., Kim, K.j., Kim, G.T., Chang, I.S., Lee, J. and Phung, N.T. (2004) Enrichment of microbial community generating electricity using a fuel-cell-type electrochemical cell. Appl. Microbiol. Biotechnol. 63(6), 672-681.

Kim, H.-J., Hyun, M.-S., Chang, I.S. and Kim, B.-H. (1999d) A microbial fuel cell type lactate biosensor using a metal-reducing bacterium, *Shewanella putrefaciens*. J. Microbiol. Biotechnol. 9(3), 365-367.

Kim, H.J., Park, H.S., Hyun, M.S., Chang, I.S., Kim, M. and Kim, B.H. (2002) A mediator-less microbial fuel cell using a metal reducing bacterium, *Shewanella putrefaciens*. Enzyme Microb. Technol. 30(2), 145-152.

Kim, J., Zuo, Y., Regan, J.M. and Logan, B.E. (2007a) Ammonia removal from wastewater enhanced by electricity generation and charge transfer in microbial fuel cells (MFCs). Biotechnol. Bioengin. Submitted.

Kim, J.R., Cheng, S., Oh, S.-E. and Logan, B.E. (2007b) Power generation using different cation, anion and ultrafiltration membranes in microbial fuel cells. Environ. Sci. Technol. 41(3), 1004-1009.

Kim, J.R., Jung, S.H., Regan, J.M. and Logan, B.E. (2007c) Electricity generation and microbial community analysis of ethanol powered microbial fuel cells. Bioresource Technol. 98(13), 2568-2577.

Kim, J.R., Min, B. and Logan, B.E. (2005) Evaluation of procedures to acclimate a microbial fuel cell for electricity production. Appl. Microbiol. Biotechnol. 68(1), 23-30.

Larminie, J. and Dicks, A. (2000) Fuel cell systems explained, John Wiley & Sons, Chichester.

Lee, J., Phung, N.T., Chang, I.S., Kim, B.H. and Sung, H.C. (2003) Use of acetate for enrichment of electrochemically active microorganisms and their 16S rDNA analyses. FEMS Microbiol. Lett. 223(2), 185-191.

Leonardo, M.R., Dailly, Y. and Clark, D.P. (1996) Role of NAD in regulating the adhE gene of *Escherichia coli*. J. Bacteriol. 178(20), 6013-6018.

Levin, D.B., Pitt, L. and Love, M. (2004) Biohydrogen production: prospects and limitations to practical applications. Int. J. Hydrogen Energy 29, 173-185.

Lewis, K. (1966) Symposium on bioelectrochemistry of microorganisms: IV. Biochemical fuel cells. Bacteriol. Rev. 30(1), 101-113.

Lewis, N.S. and Nocera, D.G. (2006) Powering the planet: chemical challenges in solar energy utilization. PNAS 103(43), 15729-15735.

Li, T. and Farrell, J. (2000) Reductive dechlorination of trichloroethene and carbon tetrachloride using iron and palladized-iron cathodes. Environ. Sci. Technol. 34(1), 173-179.

Lide, D.R. (1995) CRC handbook of chemistry and physics, CRC Press, Inc., Boca Raton.

Lies, D.P., Hernandez, M.E., Kappler, A., Mielke, R.E., Gralnick, J.A. and Newman, D.K. (2005) *Shewanella oneidenis* MR-1 uses overlapping pathways for iron reduction at a distance and by direct contact under conditions relevant for biofilms. Appl. Environ. Microbiol. 71(8), 4414-4426.

Lin, W.C., Coppi, M.V. and Lovley, D.R. (2004) *Geobacter sulfurreducens* can grow with oxygen as a terminal electron acceptor. Appl. Environ. Microbiol. 70(4), 2525-2528.

Liu, H., Cheng, S. and Logan, B.E. (2005a) Power generation in fed-batch microbial fuel cells as a function of ionic strength, temperature, and reactor configuration. Environ. Sci. Technol. 39(14), 5488-5493.

Liu, H., Cheng, S. and Logan, B.E. (2005b) Production of electricity from acetate or butyrate in a single chamber microbial fuel cell. Environ. Sci. Technol. 39(2), 658-662.

Liu, H., Cheng, S. and Logan, B.E. (2007) Scale up of a single-chamber microbial fuel cell through optimization of the anode to cathode area ratio. In preparation.
Liu, H., Grot, S. and Logan, B.E. (2005c) Electrochemically assisted microbial production of hydrogen from acetate. Environ. Sci. Technol. 39(11), 4317-4320.
Liu, H. and Logan, B.E. (2004) Electricity generation using an air-cathode single chamber microbial fuel cell in the presence and absence of a proton exchange membrane. Environ. Sci. Technol. 38(14), 4040-4046.
Liu, H., Ramnarayanan, R. and Logan, B.E. (2004) Production of electricity during wastewater treatment using a single chamber microbial fuel cell. Environ. Sci. Technol. 38(7), 2281-2285.
Logan, B.E. (1999) Environmental Transport Processes, Wiley, New York.
Logan, B.E. (2004) Extracting hydrogen and electricity from renewable resources. Environ. Sci. Technol. 38(9), 160A-167A.
Logan, B.E. (2005) Materials and configuration for scalable microbial fuel cells. Provisional patent application, PST 20918, PSU 2006-3173, Penn State University.
Logan, B.E., Aelterman, P., Hamelers, B., Rozendal, R., Schröder, U., Keller, J., Freguiac, S., Verstraete, W. and Rabaey, K. (2006) Microbial fuel cells: methodology and technology. Environ. Sci. Technol. 40(17), 5181-5192.
Logan, B.E., Cheng, S., Watson, V. and Estadt, G. (2007) Graphite fiber brush anodes for increased power production in air-cathode microbial fuel cells. Environ. Sci. Technol. 41(9), 3341-3346.
Logan, B.E. and Grot, S. (2005) A bioelectrochemically assisted microbial reactor (BEAMR) that generates hydrogen gas. Patent application 50/588,022, Penn State University.
Logan, B.E., Hermanowicz, S.W. and Parker, D.S. (1987) A fundamental model for trickling filter process design. J. Water Pollut. Control. Fed. 59(12), 1029-1042.
Logan, B.E., Murano, C., Scott, K., Gray, N.D. and Head, I.M. (2005) Electricity generation from cysteine in a microbial fuel cell. Water Res. 39(5), 942-952.
Logan, B.E., Oh, S.E., Kim, I.S. and van Ginkel, S. (2002) Biological hydrogen production measured in batch anaerobic respirometers. Environ. Sci. Technol. 36(11), 2530-2535.
Logan, B.E. and Regan, J.M. (2006a) Electricity-producing bacterial communities in microbial fuel cells. Trends Microbiol. 14(12), 512-518.
Logan, B.E. and Regan, J.M. (2006b) Microbial fuel cells—challenges and applications. Environ. Sci. Technol. 40(17), 5172-5180.
Lovley, D.R. (2006) Bug juice: harvesting electricity with microorganisms. Nature Rev. Microbiol. 4, 497-508.
Lower, S.K., Hochella, M.F. and Beveridge, T.J. (2001) Bacterial recognition of mineral surfaces: nanoscale interactions between *Shewanella* and α-FeOOH. Science 292(5520), 1360-1363.
Lowy, D.A., Tender, L.M., Zeikus, J.G., Park, D.H. and Lovley, D.R. (2006) Harvesting energy from the marine sediment-water interface II - Kinetic activity of anode materials. Biosens. Bioelectron. 21(11), 2058-2063.
Madigan, M.T. and Martinko, J.M. (2006) Brock Biology of Microorganisms, Pearson Eductation Inc., Upper Saddle River, NJ.
Matasci, R.N., Kaempfer, C. and Heidman, J.A. (1986) Full scale studies of the trickling filter/solids contact process. J. Water Pollut. Control. Fed. 58(11), 1043-1049.
Matsumoto, N., Nakasono, S., Ohmura, N. and Saiki, H. (1999) Extension of logarithmic growth of *Thiobacillus ferroixidans* by potential controlled electrochemical reduction of Fe(III). Biotechnol. Bioengin. 64(6), 716-721.
Menicucci, J.H., Beyenal, H., Marsili, E., Veluchamy, R.A., Demir, G. and Lewandowski, Z. (2006) Procedure for determining maximum sustainable power generated by microbial fuel cells. Environ. Sci. Technol. 40(3), 1062-1068.
Metcalf & Eddy Inc. (2003) Wastewater Engineering: Treatment and Reuse, McGraw-Hill, New York.
Methe, B.A., Nelson, K.E., Eisen, J.A., Paulsen, I.T., Nelson, W., Heidelberg, J.F., Wu, D., Wu, M., Ward, N., Beanan, M.J., Dodson, R.J., Madupu, R., Brinkac, L.M., Daugherty, S.C., DeBoy, R.T., Durkin, A.S., Gwinn, M., Kolonay, J.F., Sullivan, S.A., Haft, D.H., Selengut, J.,

Davidsen, T.M., Zafar, N., White, O., Tran, B., Romero, C., Forberger, H.A., Weidman, J., Khouri, H., Feldblyum, T.V., Utterback, T.R., Van Aken, S.E., Lovley, D.R. and Fraser, C.M. (2003) Genome of *Geobacter sulfurreducens*: Metal reduction in subsurface environments. Science 302(5652), 1967-1969.

Milliken, C.E. and May, H.D. (2007) Sustained generation of electricity by the spore-forming, gram-positive, *Desulfitobacterium hafniense* strain DCB2. Appl. Microbiol. Biotechnol. 73, 1180-1189.

Min, B., Cheng, S. and Logan, B.E. (2005a) Electricity generation using membrane and salt bridge microbial fuel cells. Water Res. 39(9), 1675-1686.

Min, B., Kim, J.R., Oh, S., Regan, J.M. and Logan, B.E. (2005b) Electricity generation from swine wastewater using microbial fuel cells. Water Res. 39(20), 4961-4968.

Min, B. and Logan, B.E. (2004) Continuous electricity generation from domestic wastewater and organic substrates in a flat plate microbial fuel cell. Environ. Sci. Technol. 38(21), 5809-5814.

Moon, H., Chang, I.S., Jang, J.K. and Kim, B.H. (2005) Residence time distribution in microbial fuel cell and its influence on COD removal with electricity generation. Biochem. Eng. J. 27(1), 59-65.

Moon, H., Chang, I.S. and Kim, B.H. (2006) Continuous electricity production from artificial wastewater using a mediator-less microbial fuel cell. Bioresource Technol. 97(4), 621-627.

Moraes, P.B. and Bertazzoli, R. (2005) Electrodegradation of landfill leachate in a flow electrochemical reactor. Chemosphere 58, 41-46.

Myers, C.R. and Myers, J.M. (2004) *Shewanella oneidensis* MR-1 restores menaquinone synthesis to a menaquinone-negative mutant. Appl. Environ. Microbiol. 70(9), 5415-5425.

Myers, J.M. and Myers, C.R. (1998) Isolation and sequence of *omcA*, a gene encoding a decaheme outer membrane cytochrome c of *Shewanella putrefaciens* MR-1, and detection of *omcA homologs* in other strains of *S. putrefaciens*. Biochim. Biophys. Acta 1373(1), 237-251.

Myers, J.M. and Myers, C.R. (2000) Role of the tetraheme cytochrome cymA in anaerobic electron transport in cells of *Shewanella putrefaciens* MR-1 with normal levels of menaquinone. J. Bacteriol. 182(1), 67-75.

Myers, J.M. and Myers, C.R. (2001) Role for outer membrane cytochromes OmcA and OmcB of *Shewanella putrefaciens* MR-1 in reduction of manganese dioxide. Appl. Environ. Microbiol. 67(1), 260-269.

Myers, J.M. and Myers, C.R. (2002) Genetic complementation of an outer membrane cytochrome *omcB* mutant of *Shewanella putrefaciens* MR-1 requires *omcB* plus downstream DNA. Appl. Environ. Microbiol. 68(6), 2781-2793.

Neilsen, J.L., Juretschko, S., Wagner, M. and Nielsen, P.H. (2002) Abundance and phylogenetic affiliation of iron reducers in activated sludge as assessed by fluorescence in situ hybridization and microautoradiography. Appl. Environ. Microbiol. 68(9), 4629-4636.

Newman, D.K. and Kolter, R. (2000) A role for excreted quinones in extracellular electron transfer. Nature 405(6782), 94-97.

Niessen, J., Schröder, U., Harnish, F. and Scholz, F. (2005) Gaining electricity from in situ oxidation of hydrogen produced by fermentative cellulose degradation. Lett. Appl. Microbiol. 41, 286-290.

Niessen, J., Schröder, U., Rosenbaum, M. and Scholz, F. (2004) Fluorinated polyanilines as superior materials for electrocatalytic anodes in bacterial fuel cells. Electrochem. Commun. 6(6), 571-575.

Oh, S.-E. and Logan, B.E. (2005) Hydrogen and electricity production from a food processing wastewater using fermentation and microbial fuel cell technologies. Water Res. 39(19), 4673-4682.

Oh, S.-E. and Logan, B.E. (2007) Voltage reversal during microbial fuel cell stack operation. J. Power Sour. 167(1), 11-17.

Oh, S. and Logan, B.E. (2006) Proton exchange membrane and electrode surface areas as factors that affect power generation in microbial fuel cells. Appl. Microbiol. Biotechnol. 70(2), 162-169.

Oh, S., Min, B. and Logan, B.E. (2004) Cathode performance as a factor in electricity generation in microbial fuel cells. Environ. Sci. Technol. 38(8), 4900-4904.

Pardi, S.T. (2002) Investigation of bacterial extracellular polymeric substances utilizing an atomic force microscope with a newly developed deflection curve acquisition and analysis protocol, Penn State, University Park.

Park, D.H., Kim, S.K., Shin, I.H. and Jeong, Y.J. (2000) Electricity production in biofuel cell using modified graphite electrode with neutral red. Biotechnol. Lett. 22, 1301-1304.

Park, D.H., Laivenieks, M., Guettler, M.V., Jain, M.K. and Zeikus, J.G. (1999) Microbial utilization of electrically reduced neutral red as the sole electron donor for growth and metabolite production. Appl. Environ. Microbiol. 65(7), 2912-2917.

Park, D.H. and Zeikus, J.G. (2000) Electricity generation in microbial fuel cells using neutral red as an electronophore. Appl. Environ. Microbiol. 66(4), 1292-1297.

Park, D.H. and Zeikus, J.G. (2002) Impact of electrode composition on electricity generation in a single-copartment fuel cell using *Shewanella putrefacians*. Appl. Microbiol. Biotechnol. 59, 58-61.

Park, D.H. and Zeikus, J.G. (2003) Improved fuel cell and electrode designs for producing electricity from microbial degradation. Biotechnol. Bioengin. 81(3), 348-355.

Park, H.S., Kim, B.H., Kim, H.S., Kim, H.J., Kim, G.T., Kim, M., Chang, I.S., Park, Y.K. and Chang, H.I. (2001) A novel electrochemically active and Fe(III)-reducing bacterium phylogenetically related to *Clostridium butyricum* isolated from a microbial fuel cell. Anaerobe 7(6), 297-306.

Parker, D.S. and Bratby, J.R. (2001) Review of two decades of experience with TF/SC process. J. Environ. Eng. 127(5), 380387.

Perlack, R.D., Wright, L.L., Turhollow, A., Graham, R.L., Stokes, B. and Erbach, D.C. (2005) Biomass as feedstock for a bioenergy and bioproducts industry: the technical feasibility of a billion-ton annual supply, Oak Ridge National Laboratory.

Perry, R.H. and Chilton, C.H. (1973) Chemical Engineers' Handbook, McGraw-Hill, New York.

Pham, C.A., Jung, S.J., Phung, N.T., Lee, J., Chang, I.S., Kim, B.H., Yi, H. and Chun, J. (2003) A novel electrochemically active and Fe(III)-reducing bacterium phylogenetically related to *Aeromonas hydrophila*, isolated from a microbial fuel cell. FEMS Microbiol. Lett. 223(1), 129-134.

Pham, T.H., Rabaey, K., Aelterman, P., Clauwaert, P., De Schamphelaire, L., Boon, N. and Verstraete, W. (2006) Microbial fuel cells in relation to conventional anaerobic digestion technology. Eng. Life Sci. 6(3), 285-292.

Phung, N.T., Lee, J., Kang, K.H., Chang, I.S., Gadd, G.M. and Kim, B.H. (2004) Analysis of microbial diversity in oligotrophic microbial fuel cells using 16S rDNA sequences. FEMS Microbiol. Lett. 233(1), 77-82.

Potter, M.C. (1911) Electrical effects accompanying the decomposition of organic compounds. Proc. Roy. Soc. London Ser. B 84, 260-276.

Rabaey, K., Boon, N., Hofte, M. and Verstraete, W. (2005a) Microbial phenazine production enhances electron transfer in biofuel cells. Environ. Sci. Technol. 39(9), 3401-3408.

Rabaey, K., Boon, N., Siciliano, S.D., Verhaege, M. and Verstraete, W. (2004) Biofuel cells select for microbial consortia that self-mediate electron transfer. Appl. Environ. Microbiol. 70(9), 5373-5382.

Rabaey, K., Clauwaert, P., Aelterman, P. and Verstraete, W. (2005b) Tubular microbial fuel cells for efficient electricity generation. Environ. Sci. Technol. 39(20), 8077-8082.

Rabaey, K., Lissens, G., Siciliano, S.D. and Verstraete, W. (2003) A microbial fuel cell capable of converting glucose to electricity at high rate and efficiency. Biotechnol. Lett. 25(18), 1531-1535.

Rabaey, K., Ossieur, W., Verhaege, M. and Verstraete, W. (2005c) Continuous microbial fuel cells convert carbohydrates to electricity. Wat. Sci. Technol. 52(1-2), 515-523.

Rabaey, K., van de Sompel, K., Maignien, L., Boon, N., Aelterman, P., Clauwaert, P., de Schamphelaire, L., Pham, T.H., Vermeulen, J., Verhaege, M., Lens, P. and Verstraete, W. (2006) Microbial fuel cells for sulfide removal. Environ. Sci. Technol. 40(17), 5218-5224.

Rabaey, K. and Verstraete, W. (2005) Microbial fuel cells: novel biotechnology for energy generation. Trends Biotechnol. 23(6), 291-298.

Raz, S., Jak, M.J.G., Schoonman, J. and Reiss, I. (2002) Supported mixed-gas fuel cells Solid State Ionics 149, 335-341.

Reguera, G., McCarthy, K.D., Mehta, T., Nicoll, J.S., Tuominen, M.T. and Lovley, D.R. (2005) Extracellular electron transfer via microbial nanowires. Nature 435, 1098-1101.

Reimers, C.E., Girguis, P., Stecher, H.A., Tender, L.M., Ryckelynck, N. and Whaling, P. (2006) Microbial fuel cell energy from an ocan cold seep. Geobiology 4, 123-136.

Reimers, C.E., Tender, L.M., Fertig, S. and Wang, W. (2001) Harvesting energy from the marine sediment-water interface. Environ. Sci. Technol. 35(1), 192-195.

Ren, Z., Ward, T.E. and Regan, J.M. (2007) Electricity production from cellulose in a microbial fuel cell using a defined binary culture and an undefined mixed culture. Environ. Sci. Technol. 41(13), 4781-4786.

Rezaei, F., Richard, T.L., Brennan, R. and Logan, B.E. (2007) Substrate-enhanced microbial fuel cells for improved remote power generation from sediment-based systems. Environ. Sci. Technol. 41(11), 4053-4058.

Rhoads, A., Beyenal, H. and Lewandowski, Z. (2005) Microbial fuel cell using anaerobic respiration as an anodic reaction and biomineralied manganese as a cathodic reactant. Environ. Sci. Technol. 39(2), 4666-4671.

Rifkin, J. (2002) The Hydrogen Economy, Tarcher/Putnam, New York.

Ringeisen, B.R., Henderson, E., Wu, P.K., Pietron, J., Ray, R., Little, B., Biffinger, J.C. and Jones-Meehan, J.M. (2006) High power density from a miniature microbial fuel cell using *Shewanella oneidensis* DSP10. Environ. Sci. Technol. 40(8), 2629-2634.

Rismani-Yazdi, H., Christy, A.D., Dehority, B.A., Morrison, M., Yu, Z. and Tuovinen, O.H. (2007) Electricity generation from cellulose by rumen microorganisms in microbial fuel cells. Biotechnol. Bioengin., doi 10.1002.bit.21366.

Rittmann, B.E. and McCarty, P.L. (2001) Environmental Biotechnology: Principles and Applications, McGraw-Hill, Boston.

Rosenbaum, M., Schröder, U. and Scholz, F. (2005) In situ electrooxidation of photobiological hydrogen in a photobioelectrochemical fuel cell based on *Rhodobacter sphaeroides*. Environ. Sci. Technol. 39(16), 6328-6333.

Rosenbaum, M., Zhao, F., Schröder, U. and Scholz, F. (2006) Interfacing electrocatalysis and biocatalysis with tungsten carbide: A high performance noble-metal-free microbial fuel cell. Angew. Chem. int. Ed. 45(40), 6658-6661.

Rozendal, R., Hamelers, H.V.M., Molenkamp, R.J. and Buisman, C.J.N. (2007) Performance of single chamber biocatalyzed electrolysis with different types of ion exchange membranes. Water Res. 41, 1984-1994.

Rozendal, R.A. and Buisman, C.J.N. (2005) Process for producing hydrogen. Patent WO2005005981. (WO2004NL00499 20040709).

Rozendal, R.A., Hamelers, H.V.M., Euverink, G.J.W., Metz, S.J. and Buisman, C.J.N. (2006a) Principle and perspectives of hydrogen production through biocatalyzed electrolysis. Int. J. Hydrogen Energy 31(12), 1632-1640.

Rozendal, R.A., Hamelers, H.V.V. and Guisman, C.J.N. (2006b) Effects of membrane cation transport on pH and microbial fuel cell performance. Environ. Sci. Technol. 40(17), 5206-5211.

Ryckelynck, N., Stecher, H.A.I. and Reimers, C.E. (2005) Understanding the anodic mechanism of a seafloor fuel cell: interactions between geochemistry and microbial ability. Biogeochemistry 76, 113-139.

Ryu, D.D.Y. and Nam, D.-H. (2000) Recent progress in biomolecular engineering. Biotechnology Progress 16(1), 2-16.

Schröder, U., Niessen, J. and Scholz, F. (2003) A generation of microbial fuel cells with current outputs boosted by more than one order of magnitude. Angew. Chem. Int. Ed. 42(25), 2880-2883.

Sell, D., Kramer, P. and Kreysa, G. (1989) Use of an oxygen gas diffusion cathode and a three-dimensional packed-bed anode in a bioelectrochemical fuel cell. Appl. Microbiol. Biotechnol. 31(2), 211-213.

Shantaram, A., Beyenal, H., Raajan, R., Veluchamy, A. and Lewandowski, Z. (2005) Wireless sensors powered by microbial fuel cells. Environ. Sci. Technol. 39(13), 5037-5042.

Shin, S.-H., Choi, Y., Na, S.-H., Jung, S. and Kim, S. (2006) Development of bipolor plate stack type microbial fuel cells. Bull. Korean Chem. Soc. 27(2), 281-285.

Shinnar, R. and Cintro, F. (2006) A road map to U.S. decarbonization. Science 313, 1243-1244.

Shizas, I. and Bagley, D.M. (2004) Experimental determination of energy content of unknown organics in municipal wastewater streams. J. Energy Engin. 130(2), 45-53.

Snoeyink, V.L. and Jenkins, D. (1980) Water Chemistry, John Wiley & Sons, Inc., New York.

Stams, A.J.M., de Bok, F.A.M., Plugge, C.M., van Eekert, M.H.A., Dolfing, J. and Schraa, G. (2006) Exocellular electron transfer in anaerobic microbial communities. Environ. Microbiol. 8(3), 371-382.

Tarlov, M.J. and Bowden, E.F. (1991) J. Am. Chem. Soc. 113, 1847-1849.

Tchobanoglous, G., Burton, F.L. and Stensel, H.D. (2003) Wastewater Engineering, Treatment and Reuse, Metcalf & Eddy, Inc., 4th edition, McGraw-Hill, Boston.

Tender, L.M., Reimers, C.E., Stecher, H.A., Holmes, D.E., Bond, D.R., Lowy, D.A., Pilobello, K., Fertig, S.J. and Lovley, D.R. (2002) Harnessing microbially generated power on the seafloor. Nat. Biotechnol. 20(8), 821-825.

ter Heijne, A., Hamelers, H.V.M., de Wilde, V., Rozendal, R.R. and Buisman, C.J.N. (2006) Ferric iron reduction as an alternative for platinum-based cathodes in microbial fuel cells. Environ. Sci. Technol. 40, 5200-5205.

Thauer, R.K., Jungermann, K. and Decker, K. (1977) Energy conservation in chemotrophic anaerobic bacteria. Bacteriol. Rev. 41, 100-180.

Thrash, J.C., Van Trump, J.I., Weber, K.A., Miller, E., Achenbach, L.A. and Coates, J.D. (2007) Electrochemical stimulation of microbial perchlorate reduction. Environ. Sci. Technol. 41(5), 1740-1746.

URL (2007) *E. coli* statistics, Institute for biomolecular design.

Vadillo-Rodriguez, V. and Logan, B.E. (2006) Localized attraction correlates with bacterial adhesion to glass and metal oxide substrata. Environ. Sci. Technol. 40(9), 2983-2988.

Viessman, W.J. and Hammer, M.J. (2005) Water Supply and Pollution Control, 7th edition, Pearson Prentice-Hall, Upper Saddle River, NJ.

Voggu, L., Schlag, S., Biswas, R., Rosenstein, R., Rausch, C. and Gotz, F. (2006) Microevolution of cytochrome *bd* oxidase in *Staphylococci* and its implication in resistance to respiratory toxins release by *Psuedomonas*. J. Bacteriol. 188(23), 8079-8086.

Wikipedia-Contributors (2007a) Carbon dioxide in the Earth's atmosphere [Internet], Wikipedia, The Free Encyclopedia; 2007 Jun 13, 00:55 UTC [cited 2007 Jun 13]. Available from: http://en.wikipedia.org/w/index.php?title=Carbon_dioxide_in_the_Earth%27s_atmosphere&oldid=137797293. .

Wikipedia-Contributors (2007b) Global warming [Internet], Wikipedia, The Free Encyclopedia; 2007 Jun 13, 17:54 UTC [cited 2007 Jun 13]. Available from: http://en.wikipedia.org/w/index.php?title=Global_warming&oldid=137945980. .

Willis, J., Arnett, C., Davis, S., Schettler, J., Shah, A. and Shaw, R. (2007) Maximizing methane. Wat. Environ. Technol. 19(2), 77-81.

WIN (2001) Water infrastructure now: recommendations for clean and safe water in the 21st century.

Yokoyama, H., Ohmori, H., Ishida, M., Waki, M. and Tanaka, Y. (2006) Treatment of cow-waste slurry by a microbial fuel cell and the properties of the treated slurry as a liquid manure. Animal Sci. J. 77, 634-638.

You, S., Zhao, Q., Zhang, J., Jiang, J. and Zhao, S. (2006) A microbial fuel cell using permanganate as the cathodic electron acceptor. J. Power Sour. 162, 1409-1415.

Yu, E.H., Cheng, S.C., Scott, K. and Logan, B.E. (2007) Microbial fuel cell performance with non-Pt cathode catalysts. J. Power Sour. 171(2), 275-281.

Yu, X., Amrhein, C., Deshusses, M.A. and Matsumoto, M.R. (2006) Perchlorate reduction by autotrophic bacteria in the presence of zero-valent iron. Environ. Sci. Technol. 40(4), 1328-1334.

Zhang, E., Xu, W., Diao, G. and Shuang, C. (2006a) Electricity generation from acetate and glucose by sedimentary bacterium attached to electrode in microbial-anode fuel cells. J. Power Sour. 161(2), 820-825.

Zhang, H., Bruns, M.A. and Logan, B.E. (2002) Perchlorate reduction by a novel chemolithoautotrophic hydrogen-oxidizing bacterium. Environ. Microbiol. 4(10), 570-576.

Zhang, J. and Bishop, P.L. (1994a) Density, porosity, and pore structure of biofilms. Water Res. 28(11), 2267-2277.

Zhang, T., Cui, C., Chen, S., Ai, X., Yang, H., Shen, P. and Peng, Z. (2006b) A novel mediatorless microbial fuel cell based on biocatalysis of *Escherichia coli*. Chem. Commun., 2257-2259.

Zhang, T.C. and Bishop, P.L. (1994b) Evaluation of tortuosity factors and effective diffusivities in biofilms. Water Res. 28(11), 2279-2287.

Zhao, F., Harnisch, F., Schröder, U., Scholz, F., Bogdanoff, P. and Herrmann, I. (2005) Application of pyrolysed iron (II) phthalocyanine and CoTMPP based oxygen reduction catalysts as cathode materials in microbial fuel cells. Electrochem. Commun. 7, 1405-1410.

Zuo, Y., Cheng, S., Call, D. and Logan, B.E. (2007) Tubular membrane cathodes for scalable power generation in microbial fuel cells. Environ. Sci. Technol. 41(9), 3347-3353.

Zuo, Y., Maness, P.-C. and Logan, B.E. (2005) Electricity production from steam exploded corn stover biomass. Energy & Fuels 20(4), 1716-1721.

Index